# INTERSTELLAR · TRAVEL ·

## PAST, PRESENT AND FUTURE

"It is gratifying to have an intelligent, informative overview of the whole mind-boggling subject. Macvey explores in depth a variety of techniques for overcoming the time distance barrier to interstellar travel. Throughout the book the author sustains a healthy skepticism, broad vision and playful wit." —*Publishers Weekly*

"Demanding but stimulating. He brings to his book a more solid scientific background than do some other theorists." —*Booklist*

"A distinguished astronomer, Macvey succeeds in explaining the mysteries of space in a precise, readable, plausible manner. . . . Well thought out and clearly presented." —*School Library Journal*

*Other books by the author*

*TIME TRAVEL*

*WHERE WILL WE GO WHEN THE SUN DIES?*

*COLONIZING OTHER WORLDS*

*SPACE WEAPONS/SPACE WAR*

*ALONE IN THE UNIVERSE?*

*JOURNEY TO ALPHA CENTAURI*
(published in paperback as
*HOW WE WILL REACH THE STARS*)

*WHISPERS FROM SPACE*

# INTERSTELLAR TRAVEL

## PAST, PRESENT AND FUTURE

JOHN W MACVEY
AUTHOR OF *TIME TRAVEL*

Scarborough House/*Publishers*

Scarborough House/Publishers
Chelsea, MI 48118

FIRST SCARBOROUGH HOUSE TRADE
PAPERBACK EDITION 1991

*Interstellar Travel* was originally published in hardcover
by Stein & Day/*Publishers*, and has been updated for this edition.


Printed in the United States of America

*Library of Congress Cataloging-in-Publication Data*

Macvey, John W.
Interstellar travel : past, present, and future / John W. Macvey.
—1st Scarborough House trade pbk. ed.
p. cm.
Reprint, with revisions. Originally published: Briarcliff Manor,
N.Y. : Stein and Day, c1977.
Includes bibliographical references and index.
ISBN 0-8128-8523-6
1. Manned space flight. 2. Interstellar travel. 3. Life on other
planets. I. Title.
TL793.M19 1991
629.4—dc20 90-20138
CIP

*To Margaret*

*For I dipt into the future, far as human eye could see,*
*Saw the vision of the world, and all the wonder that would be;*
*Saw the heavens fill with commerce, argosies of magic sails,*
*Pilots of the purple twilight, dropping down with costly bales.*

ALFRED, LORD TENNYSON

# CONTENTS

# ACKNOWLEDGMENTS

A book of this nature can hardly be the result of the author's endeavors alone. In particular I wish to acknowledge the considerable assistance accorded me by Miss Helen Campbell, who is responsible for producing quickly and efficiently the many essential and sometimes complicated diagrams.

I must also thank Miss J. Kinniburgh for her expeditious typing of the script while coping nobly with frequent changes and additions on the part of the author.

I also acknowledge with deep gratitude the invaluable and much appreciated assistance accorded me within the United States by my friend, Mr. M. K. Parkhurst of New York.

It is fitting also that due tribute be paid to the publishers, editor, and editorial staff for their interest, support, and friendly guidance throughout the preparation of the book.

To all others who by virtue of their faith and encouragement have helped this book along the way go my sincere and heartfelt thanks.

*Saltcoats, Scotland* JOHN W. MACVEY

# PROLOGUE

---

All this visible universe is not unique in nature and we must believe that there are, in other regions of space, other worlds, other beings and other men.

LUCRETIUS, 99–55 B.C.

There were men from the sky in the Earth in these days.

HEBREW BOOK OF LIGHT, 12th–13th century

Innumerable suns exist; innumerable Earths revolve around these —living beings inhabit these worlds.

GIORDANO BRUNO, 1548–1600

Highly developed civilizations on planets of the suns in our galaxy should be the rule rather than the exception.

DR. JAN GADOMSKI, Polish astronomer

There are many abodes of intelligent life in the universe and many of them are inhabited by organisms with intelligence far higher than ours.

DR. LEWIS W. BECK, University of Rochester

Life on Earth may have started when spacemen landed here billions of years ago.

PROF. THOMAS GOLD, Cornell University

Intelligent beings abound in the universe and most of them are far older than we are.

DR. W. HOWARD, Harvard University

The number of inhabitable systems is about 3 to 5 per cent of the number of stars; this leads to eight billion inhabitable systems in our galaxy.

DR. SU-SHU HUANG, Dearborn Observatory

In the distant future we will encounter some other intelligent life.

FRANK BORMANN, astronaut

Something unknown to our understanding is visiting this Earth.

DR. MITROVAN ZVEREV, Soviet scientist

U.F.O.'s really exist and apparently come from other planets.

JAVIER GARZON, National Astronomical Observatory, Mexico City

The hour is very late, and no one can guess how many strange eyes and minds are already turned upon the planet Earth.

ARTHUR C. CLARKE, science writer

# PREFACE

Over the past two decades the subject of extraterrestrial life has gradually changed from a "way-out" science fiction concept to one of virtual respectability. This is heartening, for it would seem to indicate that, in some spheres of human endeavor at least, the rule of the closed and inflexible mind is coming to an end.

There can be little doubt that this change has been brought about largely by mankind's first feeble and faltering footsteps in space and perhaps to a lesser extent by his realization that the planet Earth is only a very minute fragment of the universe—a mere super-spaceship carrying the human race and a multitude of other creatures on a long and at times arduous journey through space and time.

Today it is apparent to most rational minds that we can hardly be alone in this vast and virtually limitless universe of ours. Cosmology has shown us how our planetary system was created and evolved. It has also shown us that our sun is but a stable, average star and that, far from being unique, planetary systems are in all probability extremely common.

In their turn biologists, though still perhaps unable to pinpoint precisely the initiators of life, have demonstrated by experiment a number of eminently reasonable ways by which the inert could have been transformed into the living. In the circumstances it would seem quite ridiculous to assume that only one small planet, our own, could be unique in this respect.

We see therefore how we live in a great galaxy containing some 100,000 million stars, a fair proportion of which almost certainly possess planetary systems. On the third planet of an average yellow star some two-thirds of the way from the center of the galaxy life was somehow initiated; it evolved and became, technologically at least, reasonably intelligent. What took place on Earth must also have taken place at countless other points within the galaxy—and indeed within all the other galaxies.

The inhabitants of the third planet of an average star have already begun to explore their cosmic "inland sea," the sister planets of Earth. Sooner or later they will make their first hesitant probings into the great interstellar ocean and, as their technology develops, begin to plan the first voyages to the nearer stars.

At this point, if indeed it has not already done so, another thought begins to intrude. What of intelligent beings already sufficiently advanced not only to make short interstar journeys but also eminently capable of interstellar exploration and colonization on a grand scale? What if our sun and solar system are destined to be the focus of such expeditions?

Of course, it has happened already in the pages of science fiction. The late H. G. Wells probably started it back in 1897 when he launched his octopuslike Martians at us. At that time, and for a couple of decades thereafter, science saw Mars as a planet probably inhabited by a race of master scientists, engineers, and technologists. Why should the writers look any farther for BEMs ("bug-eyed monsters") and such? Since then, however, in fiction Earth has been invaded, infiltrated, fought over, poisoned, intimidated, and destroyed by about every conceivable type of life from the nearest star to the remotest galaxy.

Despite the possible profusion of life within our galaxy it must of necessity be well scattered. The average distance between stars is probably of the order of 5 to 10 light-years; that between intelligent races having advanced space potential as much as 300 light-years.

Recently there has been a spate of books seeking to further the idea that at intervals throughout our world's long past alien beings from the stars have visited us. Many examples have been cited of artifacts here on Earth which may not after all be of indigenous origin. In some instances there does seem a possible validity for such a claim but in most it is difficult to avoid the conclusion that the writers are desperately and passionately trying to equate fact with theory. Invariably, too, they wholly gloss over the means of alien transit from the stars. Readers are simply supposed to assume that the aliens had this not inconsiderable power.

This seems less than honest. If aliens have of necessity to travel upward of 300 light-years to reach us and do so well within their lifetimes, then clearly they must—and about this there can be no argument—have highly sophisticated and advanced transit systems and forms of propulsion. If we are to accept the concept of possible visits, past, present, or future, then we simply *must* concede this. First, therefore, we must endeavor to ascertain whether or not there could be viable methods.

Now of course there is an immediate and very obvious difficulty here, for what we are proposing is nothing less than an attempt to examine techniques which, so far as we here on Earth are concerned, could lie something like a thousand years or more in the future. In these circumstances we can only speculate. Certainly we cannot possibly come down heavily in favor of any particular technique. What we must do—all in fact we *can* do at present—is to consider possible avenues of approach. We may at times be on the right track; other times we may not. Indeed, we might possibly omit the very technique that has given alien beings the key to the universe.

It is also fitting in the circumstances that we think a little about possible alien form. At present, especially in the realms of science fiction, the tendency is to elect for one of two extremes: the extremely repulsive "bug-eyed monster" or the blond, statuesque, godlike creature from the stars. Common sense would seem to indicate some degree of compromise. Let us consider very briefly a few alternative and perhaps viable biochemistries. If we have been visited, are being visited, or are going to be visited, then let us at least have some notion of the possible form of our visitors.

With these thoughts very much in mind, my aim has been to make this book different from the others. We will first of all look at the possibilities and potential of deep and sophisticated space travel and endeavor also to gain a little insight into the form our alien visitors might assume. Then, and only then, will we examine possible instances and probable evidence of extraterrestrial visits past and present.

Such a book is speculative. By its very nature it must be. For this the writer makes no apologies. Science is fact. It also implies a willingness to retain an open mind. If not, it atrophies. Too much speculation is inadvisable. Totally closed or reluctant minds are even more so. There has to be a middle path.

Our great galaxy is vast, yet it is but one of a countless host. Life there must surely be at many points in its starry depths, life which on the wings of technology has already traveled far. Earth, our own small, familiar world, may not forever remain a cosmic backwater. If a ship (or ships) from the stars has not already visited us, then the day could yet dawn when one shall. Before we reach the stars, perhaps long before, the stars may reach us! It would be as well to have given the matter a little thought while there is yet time.

*John W. Macvey*

# PART I

# 1. THE EDGE OF BEYOND

As a prelude to the ensuing pages I would earnestly ask the reader to select a clear and moonless night, preferably in late autumn, and having done so to wander out some way into the open countryside far from the bustle, noise, and light of city or town. Above will be found one of the most enthralling and beautiful spectacles that nature can provide. As a gentle breeze rustles and sighs over the dark surface of our world the eternal heavens straddle it with a panoply of exquisite, incredible splendor. There is a feeling of infinite peace. The prosaic, mundane affairs of mortal men are seen in their proper context—absurd, irrelevant, and inconsequential. For a countless million years, long before man or his ancestors were fashioned, these stars were there and they will still be there long after man has vanished from the terrestrial scene. In the mystery of these depths is continuity from alpha to omega, the infinite both in space and time.

There is too another highly intriguing aspect, an aspect regarded not so long ago as fantastic, one certainly less than respectable in the eyes of established science of that period. Times, however, have changed, and are indeed still changing. We are members of a race spawned on a tiny planet of an undistinguished and very average star. As we stand silently under the celestial glory of the night sky how impossible it is to believe that we can possibly be alone in this infinity. Each little glittering pinpoint of light in the nocturnal dome above is another sun, many almost certainly with planets too, and on a proportion of these, other peoples and other civilizations. At a myriad of points within that sweeping, star-powdered vault other intelligent beings must even now be gazing skyward, seeing the same celestial glory. One of the stars some must see will be our own familiar sun. If it shines brightly in their skies it will almost certainly be the subject of special scrutiny. Alien minds will ponder on whether or not it has planets and wonder if thereon intelligent beings look outward toward them.

It is also extremely interesting to reflect not just on the stars themselves but also on the dark spaces between them; spaces which the strange geometry of space renders so small yet which we know to be gulfs of unbelievable and frightening immensity.

Already from Earth we have sent craft and men to traverse the very short gap between our world and the moon as well as unmanned probes to the nearer (and by now some of the not-so-near) planets. Indeed, one such probe is even now about to leave the solar system forever on a journey which several millennia hence will find it among some of the stars of the constellation Taurus. Have beings with massive intellects, thousands of years in advance of our own, sent machines and fellow creatures not just into the regions of their own planets but into the abysses which separate sun from sun? It is a thought which cannot and should not be any longer dismissed.

And so as we gaze into impalpable depths we can also contemplate the possibility of mysterious alien vessels traversing them. It is an uncanny, sobering thought. Strange beings in strange vessels up there, in a sense surrounding us. How different will they be? We may reflect on what odd dramas are being enacted, what achievements, what sagas, what tragedies. We may even think of the hopes and fears, aspirations and doubts, cruelties and kindnesses, hate and love.

What types of space vessels will these be and is it possible for us at this point in time to visualize something of their form and the nature of the forces that propel them? This is a question which we can approach in two ways. The first and most obvious of these is to regard interstellar travel merely as an extension of techniques already used in interplanetary travel. Such extensions are more in the nature of extrapolations since they must allow for parameters of time and distance of vast and unprecedented dimensions. These include generation travel (space arks) and the use of cryogenics (suspended animation). What we are seeing here is the superimposing of new facets on the orthodox techniques of interplanetary travel. We are thereby confronted with the concept of very large starships moving at relatively low velocities and therefore taking long periods to reach their stellar destinations.

It may well be that other races in the universe technologically ahead of us by only a century or two have already adopted such practices, from which it follows that somewhere within the impalpable depths of the night sky vessels even now may be striving to cross the great divides bearing either successive generations of male and female astronauts or inert travelers in a sleep spanning centuries. Fascinating though these concepts are, it is no part of our plan to enlarge upon them here.

The second approach is to think in terms of rapid interstellar travel. This invokes techniques which we must of necessity attribute to highly intelligent beings millennia ahead of us. Not for such races a slow crawl between stars lasting centuries but swift, effective transit involving only weeks or months, perhaps even less. Is it possible that techniques could be developed rendering such a thing possible? At this juncture in human affairs it is a question which clearly we cannot hope to answer.

It might be argued, and no doubt a good case be made, that such a thing is indeed impossible by *any* standards. This could conceivably be so, yet much less than a hundred years ago the prediction that by now twelve men would have walked upon the moon would have been received with total incredulity. It may be said that comparisons of this nature are invidious and do not constitute valid reasoning. Nevertheless, the facts must be allowed to speak. To Stone Age man a journey to the moon would have seemed every bit as fantastic as does the vision of interstellar travel today. We are still in the shallow, fringe waters of physical and cosmological understanding and must therefore accept it as a distinct possibility. Consequently we must suppose that highly intelligent beings in other parts of the galaxy will have discovered the underlying principles and developed the necessary technology. This being so, then in the starry heavens above and surrounding us strange, bizarre craft even now flit quickly and effortlessly from star to star. Will eventually our sun be their destination star, or are they already looking us over?

This, then, is our theme; to investigate and speculate, remembering at all times the great limitations which the present scanty state of our knowledge imposes. We will be peering myopically into the future searching for roads which, even if we found them, would be impossible as yet for us to follow.

The late Albert Einstein lit a torch. Its rays have shed a little light but already they have revealed some strange and unsuspected things. Our universe is not the logical, cut-and-dried universe of Newton. It is an odd place, and the more closely we look the odder it seems to become. We are beginning to realize that we barely understand it at all. There may exist roads to the stars undreamed of by us—strange, unexpected bypasses rather than the long, arduous trails we presently accept as inevitable. *We* certainly cannot hope to pass along these sophisticated "tunnels" in space, but other beings may. If so, how often has our sun lain at such a "tunnel mouth"? Perhaps *as yet* never; perhaps more frequently than we think!

# 2. STARS, MEN, AND ALIENS

Before looking into the possibility of alien visitations to our planet, past, present, and future, it is desirable to review the fundamentals of transit between the stars. It is not, however, the purpose of this book (and this cannot be too strongly emphasized) to deal with techniques representing mere extrapolations of present-day thinking and technology. Our aim here will be to examine and scrutinize evidence both past and present to support the growing contention that visits by representatives of technologically advanced galactic races cannot be ruled out and to dwell, so far as is currently possible in the light of late-twentieth-century science, on the bizarre and ultrasophisticated techniques that beings several millennia ahead of us might have adopted in order to achieve this end. So far as the latter aim is concerned, the difficulties are so obvious that they speak rather effectively for themselves. Nevertheless, it is essential that the attempt should be made. Fortunately there are one or two avenues which we can conveniently and perhaps profitably explore.

First, it is desirable to reiterate a number of fundamentals concerning the universe itself and to consider the implications arising from these.

Most of us are already aware that our life-giving and vitally essential sun is merely another star and a very average and undistinguished one at that. It has a family of nine planets (of which Earth is one of the least impressive), about forty moons, a host of asteroids or minor planets, an unknown number of comets, and an infinity of small particles known as meteors. The solar system looked at from a purely terrestrial standpoint appears almost incomprehensibly vast. Despite a high degree of optimism during the closing years of the last century, life (certainly advanced, intelligent life) seems to be confined to only one body within it. That body, needless to say, is our own familiar planet.

It would, however, be inexcusably and unreasonably parochial to consider life in the universe merely in the context of our solar system for the solar system is but an infinitesimal fragment of the universe. We must survey the whole immense panorama of space in which our sun and system of planets, moons, asteroids, and comets are as grains of sand on a beach. The sun is a star. Conversely all other stars are suns—suns that are incredibly remote by terrestrial standards, suns that must almost certainly in many instances be the possessors of planetary families. Because of that remoteness the most powerful telescopes yet constructed are quite unable to reveal these strange other worlds.

Recent estimates based on sound astrophysical reasoning put the proportion of stars likely to have a planetary retinue at around 70 percent.[1] Of these, probably 10 percent have habitable planets. Since the stars in our galaxy (the Milky Way) number about 135,000 million there may therefore be about 87,500 million planetary systems, 8,750 million of which may possess habitable worlds. If such systems approximate to our own (and we have good reason to believe that they do), then perhaps only one planet in each system will be habitable. Thus our galaxy could easily contain 8,750 million inhabited planets. If we decide to err on the safe side (never a bad idea in considering questions like this) and regard only a tenth of this figure as acceptable, then we still have the possibility of over 800 million planets containing life of one sort or another. Now let us suppose that of these only 10 percent contain life-forms that are well in advance of our own technologically—result, some 80 million planets containing beings likely to have perfected realistic techniques for crossing the yawning gulfs of interstellar space.

Were these planets and their parent stars all located in a single cluster around our sun we would in all probability not have to debate the pros and cons of extraterrestrial life. By now we would probably be all too uncomfortably aware of it! These 80 million or so planets are, however, much more likely to be distributed throughout the Milky Way though not necessarily in a uniform manner. Because of the immensity of the Milky Way only a minute fraction of these worlds can be regarded as "close" to us in terms of "conventional" interstellar propulsion techniques.

The overriding factor is distance—distance of such magnitude that the supposedly "high" velocities that have taken men to the moon and instrumented probes to the environs of Neptune would be quite useless with respect to a journey even to the *nearest* star unless our lifetimes were of the order of many thousands of years.

It is easy to quote actual distances—to say that the moon is a quarter of a million miles distant, that Mars at its closest is some 35 million miles remote, that Alpha Centauri, the nearest star, lies 4.3 light-years from us (25 million, million miles). Such figures mean very little, for distances of these magnitudes simply cannot be visualized. Let us instead use a scale of one foot to every million miles. On such a scale the moon is only 3 inches from us, Mars (at its closest) 35 feet, and Pluto, the outermost planet of the solar system, almost 1,200 feet, or two-thirds of a mile. On the same scale the nearest star is 5,000 *miles* distant—and it does not follow that an advanced technological society has developed there. Ten thousand miles on our scale, and we could still find ourselves alone. Another 90,000 scale miles and perhaps only then would we come upon another galactic community. If we convert this to actual distance we find that it amounts to 90 light-years or, if we insist on familiar terrestrial units of distance, 500 million, million miles. By our standards these are vast, incomprehensible distances, but by galactic standards they are relatively short. Such is the peculiar geometry of space. It is not something to which the human mind readily adapts, if at all.

About half a century ago most astronomers and cosmologists believed our sun to be rather unique, one of the very few stars in the galaxy likely to possess a planetary system, if not the only one. Consequently, so it was supposed, our race must be equally unique, one of the few, perhaps the only one in the galaxy. Looking back with the benefit of hindsight and the aid of today's cosmology such a belief sounds strangely naïve. It might even be described as arrogant and prejudiced.

The underlying reason was, of course, the idea that only the close approach to our sun by another star could have produced the heterogeneous assembly of bodies which we know as the solar system. This was the renowned "Tidal Theory" which, with a number of variants, held almost undisputed sway for several decades. Had such a theory been tenable, our solar system would indeed have been unique, for so remote are stars from one another that the odds against two colliding or even making a close approach are tremendous. On reflection this is hardly a cause for regret!

The Tidal Theory was intriguing, colorful, and in a superficial sort of way did at the time seem to explain the existence of the sun's family of worlds rather well. Unfortunately for its protagonists it could not stand up to close analytical examination and eventually it was superseded by a theory bearing a relationship to that which the other had earlier displaced, a somewhat ironic state of affairs in the circumstances.

In its essentials we now see the birth of the solar system (and that of any other planetary system) as an event likely to occur during the lifetime of many stars, probably even of the majority. In other words, it is possible for an isolated star to produce planets without the intervention of another. Whether the precise mechanism involves a star shedding concentric shells or rings of matter, which later condense into planets, or planets being created coincidentally with the star may still be a matter for conjecture and debate. In our context this does not greatly matter. What primarily interests us is the fact that it *is* possible for stars *other* than the sun to possess planetary families. In the absence of such planetary families it is obvious there could be no alien peoples and therefore no alien starships to visit Earth or anywhere else.

So much for theories. Unfortunately, postulating the existence of other solar systems still leaves lingering doubts. Such planets must be so remote that irrespective of the degree of optical aid employed it is impossible to discern them. It must always be remembered that planets in relation to their parent stars are very small, they do not shine by virtue of their own light, and by the standards of interstellar distance they lie very close to their parent stars.

Had there been no other way in which to confirm the existence of other planetary systems the position for the foreseeable future would no doubt have had to remain very much as it was. Fortunately this was not so. Stars from year to year, indeed from century to century and longer, appear, with one or two notable exceptions, to remain fixed in the sky relative to one another. For this reason they are termed "fixed stars" by astronomers. But in fact each star possesses a proper motion, that is, it is steering a course through the galaxy at a very high velocity. Our own sun is certainly no exception. At a velocity of 12 miles per second it is moving steadily toward a point in the sky close to that presently occupied by the lovely summer star Vega in the constellation of the Lyre. Fortunately for us, Earth, along with all the other planets and bodies of the solar system, is going along with it! It is only the effect of sheer distance that makes the stars seem immobile. A useful analogy to illustrate this point is the case of the bee and the jet plane. A bee buzzes past our head only a few inches from us. It appears to be traveling very fast, whereas the jet flying overhead at 30,000 feet barely seems to be moving. We know very well of course that the bee has not gone supersonic, while the high-flying jet may well be exceeding the speed of sound. It is all very much a question of distance. The bee is close to us; the jet is not.

Specialized instruments and techniques render it possible to reveal a star's true motion. They are able to reveal even more than that. A star having a large, massive planet among its retinue will deviate very slightly from a straight course by virtue of the gravitational effect of that planet. The star will continue to travel in the same direction but superimposed upon its straight-line course will be a slight yet none the less distinct "wobble."

In 1942 a minute deviation of this type was observed in the course of the binary (double) star system of 61 Cygni, a deviation which could be explained only by postulating the presence of a large orbiting planet (or planets) in its immediate vicinity. Though such a planet represents a most unlikely locale for life of any description, it must always be borne in mind that where large, massive planets exist so also may smaller ones, for example, huge Jupiter and tiny Earth in our own solar system. Small terrestrial-type planets will of course produce such infinitesimally small deviations to a star's proper motion as to be indiscernible, especially from distances of an interstellar order.

Since 1942 a similar state of affairs has been discovered at other widely separated points in the heavens. A very large, massive planet is strongly suspected in the environs of the star 70 Ophiuchi, while in 1963 the existence of a planetary companion or companions to Barnard's Star was claimed by the eminent astronomer Peter Van de Kamp.

A second measure of direct proof for the existence of extra-solar planets lies in the uneven distribution of angular momentum within the solar system itself. Whereas the bulk of our system's mass (99.9 percent) is contained within the sun (as we might reasonably expect) nearly all the system's spin (98 percent) is associated with the planets (which we would certainly *not* expect). There appears to exist, therefore, a strong case for supposing that the primordial gaseous nebula which spawned the sun must have been robbed of its spin or angular momentum by the creation of planets. In fact, a fairly recent estimate would seem to indicate that had the sun been minus its planets it would presently be rotating fifty times as fast.

Thus, alien astronomers studying our sun might be inclined to regard it as planetless on account of the lack of appreciable "wobble" as it journeys through space. It is unlikely that Jupiter, the largest planet in the solar system, is sufficiently massive to cause a high enough degree of "wobble." On the other hand, the sun's relatively slow spin would, to their astrophysicists, be strongly indicative of the birth and subsequent development of a planetary family.

Spectroscopic examination of the stars has produced a somewhat revealing state of affairs, that is, that whereas young and very hot stars possess a high degree of rotation, older, cooler ones have markedly less. The inference is that in producing planets a star loses a great deal of its spin or angular momentum to these planets. On this basis the proportion of stars likely to possess planets is around 67 percent.

This does not of course guarantee that all these possible planets are appropriate to the initiation and development of life. Though we may certainly extrapolate to a limited extent the conditions for life processes, it should be obvious that this procedure cannot and must not be carried to ridiculous extremes. The essential biochemistry of extraterrestrial life may not always be akin to our own. Neither, on the other hand, can it be too fundamentally different, for the basic laws of biology, chemistry, and physics must still obtain. If some science fiction writers had kept a slightly tighter hold on the reins in this respect we might perhaps have been treated to fewer improbable little green men and equally improbable bug-eyed monsters!

The biochemical aspects of the question are extremely intriguing and in Chapter 8 we will be taking a slightly closer look at this aspect.

For the present it might be a good idea to take a quick look at interstellar travel and its implications from the point of view of conventional and orthodox space technology. Having done so we should be better able to appreciate the very awesome problems it poses and the reasons why advanced alien communities must have sought more viable and sophisticated techniques in order to roam the galaxy at will.

Realizing the all too obvious limitations of the chemical rocket so far as the parameters of star travel are concerned, the obvious first alternative is to consider the possibilities of some form of nuclear propulsion. A fusion reactor using the isotopes of hydrogen, deuterium, or tritium is a likely prospect. Since tritium is radioactive with a half-life of approximately twelve years it cannot be stored for protracted periods of time.

The average distance between stars in our part of the galaxy is of the order of 5 to 10 light-years. In the case of a starship powered by a reactor of the type just mentioned minimum transit time is estimated to be around 14 to 15 years *per light-year.* Consequently, transit times for fusion-powered ships are going to be about a century to the nearest stars, which in view of the life-span of terrestrial man is hardly a viable proposition. The lifetimes of certain alien beings could be appreciably longer than our own but even if these were two to three times as great a century spent on a one-way journey would surely represent too high a proportion to be acceptable. This assumes of course that such beings reason as we do and see things in the same perspective. A tremendous dedication to reaching and exploring other star worlds could presumably change this. With fanaticism as the impelling urge, sacrifice on this scale might be regarded as a very low price to pay.

So far as we can tell, then, fusion drives could permit interstellar transit times of the order of a century in the case of the nearer stars. Were fusion drive to represent the ultimate in propulsion technique for us, as well as for aliens, manned interstellar flight could probably never be made for purposes other than colonization. For purely scientific missions the use of automated, unmanned probes would probably be essential. In Chapter 9 reference will be made to a recent and rather remarkable suggestion that already there may exist within our solar system such a probe originating from a planet of the star Epsilon Boötis.[2] There seems no valid reason why an alien probe should not one day come our way—or have already done so! Of course the mathematical odds against this tend to be long. Our sun is but one star, one of an immense host. From a "close" star such as Alpha Centauri a probe could well be sent in our direction, but the chances of a visitation by a wandering, undirected probe are considerably more remote. Neither we nor any of our galactic neighbors could utilize fusion drives of the type already mentioned for missions of exploration, scheduled passenger services, or routine freight flights, and it is certainly difficult in the light of our present and foreseeable future technology to see how enhanced fusion propulsion systems, photon drives, or the like could materially alter the unsatisfactory time factor. This is perhaps unfortunate since many of us have over the years derived considerable pleasure and escapism from reading the vivid accounts of interstar travel provided by several generations of science fiction writers. The ingredients by now are all very familiar—sophisticated starliners, the jet set of the future, high living in deep space, attractive stewardesses, glamorous heroines, rugged, daring captains, all the glitter, promise, and adventure of a new golden age. We may however be consoled, for all these things may yet come, though in ships of a very different kind traversing a medium that is frighteningly different.

Our sun is a member of a *small local cluster* of stars within the Milky Way. Recent estimates of the number of stars in that cluster having habitable planets is around two hundred. These could probably be colonized successfully by us if we had the means of reaching them and of course equally well by aliens possessing the necessary capabilities. Because of the presumed more advanced techniques of aliens certain essentially unsuitable planets might even be rendered habitable. This could hardly apply to large, gaseous, high-gravity planets akin to Jupiter, Saturn, Uranus, or Neptune but might be valid in the case of those like Mars. Aliens might therefore have a very definite and real purpose in roaming the galaxy. Certainly such "environmentalized" worlds would be costly and difficult to maintain. It is essential, however, to view the scene from alien eyes—eyes which could for long have surveyed with increasing disquiet and alarm dying home planets. Survival is a powerful force. From our point of view this can hardly be the most pleasant of prospects, reminding us, as it must, of many of the horrific aspects of H. G. Wells's famous *War of the Worlds.* Readers of this enthralling and timeless work will no doubt recall only too well the utter ruthlessness and single-mindedness of the invading Martians toward the unfortunate people of Earth. (Those who are able to recall the now legendary Orson Welles radio dramatization of the book in America one evening in October 1938 will need no further reminding of the feelings and emotions engendered!)

For a little longer let us continue to think in terms of fusion-powered starships, assuming, though only for the present, that nothing better is feasible and that the peculiar, unorthodox techniques discussed in later chapters can exist only as pipedreams. How might advanced aliens deal with this particular situation?

It could be that at a few points within our "local" cluster of stars (and almost certainly at many within the galaxy as a whole) there exist races capable of producing genetic mutations in a controlled manner. On the great majority of possible planets, however, it is also virtually certain that no aliens, irrespective of their forms and metabolisms, could survive unprotected even allowing for any measures of genetic mutation achieved. Such planets are simply inimical to life, any life! Thus if aliens are to live in a controlled environment that environment might as well approximate as much as possible to their own.

Certain environmental features must almost certainly remain uncontrollable. Probably the most notable in this respect is gravity. Since this cannot be changed, alien colonists destined for a planet of gravity higher than their own might in time be genetically adapted to cope with this. However, this could take place only within certain narrow, well-defined limits.

Suppose for the sake of argument that each and every alien colony in the galaxy is capable of establishing just *one* new colony in another star system. By virtue of such a continuing process aliens could ultimately reach the farthest limits of the galaxy. A long process? Undoubtedly so, but if such a process has been proceeding for countless millennia (as it could), then by now it is well advanced. What then of the as yet (apparently) inviolate solar system? An excuse is not necessarily being sought for UFOs, "flying saucers," and all other forms of spaceborne hardware. On the other hand, such possibilities cannot be overlooked.

Fusion-powered starships launched from stars "near" our sun could, it is reckoned, reach the far rim of the galaxy in something like a million years, a distance of 80,000 light-years. Colonization spreading gradually from star to star would obviously be a much slower process. Nevertheless, the entire galaxy could probably be colonized in something of the order of 10 million years. On the vast geological time scale of Earth this represents a very short period and is only a fraction of the time that man has existed on our planet. Galactic colonization could thus have begun while we were still evolving from our apelike ancestors. It must never be forgotten that there are a reasonable number of conveniently placed suns, a proportion of which are much older and maturer than our own. *Is a wave of remorseless, alien colonization reaching out toward us even now?* Throughout the centuries of recorded history there are many records of odd manifestations in our skies. The flying saucer legend is comparatively new. That of the unidentified flying object certainly is not!

Let us consider for a moment a homely parallel. The Americas were not discovered in a day. Whatever the claims of Columbus, his contemporaries, and near-contemporaries, it is by now reasonably well established that ships from northern Europe made some sort of landfall in the Americas centuries before. Columbus in a sense could be said to represent the vanguard of the main force whereas the earlier ships, however unintentionally, were fulfilling a probing, reconnaissance role. The parallel is plain. If alien spacecraft from the stars have made a landfall on Earth in the past they could represent the advance scouts, forerunners of a vanguard the like of which we dare hardly imagine. Our galaxy is some 10,000 million years old and contains something on the order of 100,000 million stars. It is hard to believe that in all this myriad host of suns and throughout all these countless millennia not one race has developed without achieving a degree of practical interstellar capability.

There exists another quite fascinating line of thought. Has in fact the solar system already been colonized in the past? As we will see in the second part of this book no completely unequivocal evidence of such colonization, even of brief, desultory visits, has been found so far. This is hardly surprising in the circumstances. Take, for example, our own indigenous past civilizations. How much today do we really know of them—or more correctly how little? A classic case is probably that of the Hittite Empire. Around 1500 B.C., less than three and a half thousand years ago, it embraced all of Asia Minor and spilled over into large areas of the present-day Middle East. Along with Egypt and Mesopotamia it constituted one of the principal powers in that region, yet it was only during the nineteenth century that mankind became aware of its existence. Our late-twentieth-century civilization is infinitely more complex, grandiose, and far-flung, yet were it terminated abruptly today or tomorrow (and there are times when this seems not altogether improbable!) what traces would remain after a few million years? Very little, we suspect.

The remains of former alien colonies are much more likely to be found on worlds which are airless and where the geology has been considerably less turbulent. In this respect we might think of the moon (though there the geology was turbulent too but in a different way), some of the satellites of Jupiter and Saturn, and perhaps even Mars. So far, it may be argued, no such evidence has been forthcoming from the moon. At this juncture it might be a mistake to rule out this possibility entirely. Admittedly, six successful manned expeditions have landed on its surface, twelve men have left their footprints on the dust enshrouding it. Nevertheless, the area covered by these twelve men and by the few American and Russian probes constitutes a very small percentage indeed of the moon's total surface area. Our satellite may yet spring a few surprises on us. The impact and import of such a discovery would be immeasurable. It requires only the finding of some small and trivial artifact, or even just part of one, some trifling object bearing a few unknown and indecipherable hieroglyphics, and the world would know beyond all doubt what so many have for long suspected, that in this vast universe man is not alone.

From time to time there have been suggestions that our race is not indigenous to this planet, that its origins ought to be sought elsewhere. There are two distinct variants to this theme: either that our race represents the descendants of alien colonists or, alternatively, that the spark of life was inadvertently kindled on Earth by space visitors where through countless ages it grew and developed.

Without doubt these are highly intriguing thoughts but at the present time only a very few are willing to give them credence. The evidence that man and all the other vertebrate and primate creatures on our planet are indigenous seems sufficiently conclusive to place the matter beyond any reasonable doubt. There remains, however, the remote possibility that certain species of our flora and fauna were implanted on Earth by cosmic beings aeons ago.

It has been suggested that alien cultures able to create societies with life-spans of very considerable duration would in the fullness of time be able to reach and perhaps colonize every planetary system in the galaxy. In so doing they would eventually oust all other species having shorter-lived societies and become in the end unique—the master race of the galaxy. Statistically this does not appear too unreasonable but in practice it is probably unrealistic. Such a society would require a life-span having a duration equal almost to that of the galaxy itself. We are aware that stars and therefore galaxies exist for incredibly long periods. We also know that a considerable portion of a star's life has passed before even the most primitive of life-forms evolve from primeval chemical "broths." Vast intervals are required also for the development of these. Stars in each and every galaxy have their allotted life-spans. When they reach the end of these they must snuff out very effectively and permanently all life on any planets surrounding them. We cannot therefore logically envisage societies enduring for periods comparable to that of galaxies.

During the last decade or so much interest has been aroused in the concept of interstellar communication by electronic means. The possibilities in this field are by no means inconsiderable and the future (not necessarily the remote future either) may hold many surprises. Interstellar radio transmissions meant for other ears might be intercepted by us, though at present the odds are against this. A recent mathematical analysis of the possibilities by Sebastian von Hoerner substantiates this. (We must remember that in the future and, in fact, even now, some such signals might be intended specifically for us.) In these circumstances we should not lay too much stress or emphasis on an absence of interstellar signals. This might be only a seeming absence, an illusion. We would certainly be unwise to accept this as proof of the nonexistence of fellow cosmic beings.

In terms, then, of conventional starships (and in this context it is surely valid to regard fusion propulsion as conventional) it would be surprising to see alien starships spiraling down through our skies unless the creators of such vessels had already established a colony or at least an advanced base somewhere within the confines of the solar system. But surely, the reader is entitled to ask, we could reasonably expect to detect radio traffic from sources relatively so near to us? This is not necessarily so. The distances involved, though far removed from light-year order, are considerable. Pluto, outermost planet of the solar system, is 5½ light-*hours* distant and the large moons of the other outer planets are also very remote by terrestrial standards. Even Mars at its nearest (35 million miles) is not exactly on our doorstep. Radio traffic from alien sources *within* the solar system could be of a type not intended for our ears. It would not be beamed in our direction and so would not reach us—reach us, that is, in detectable strength. Amid all the clamor, static, and general background noise of the airwaves it would very likely be lost.

What, however, of alien ships in the close vicinity of Earth? This is obviously a different matter. An alien colony on the moon would almost certainly give itself away electronically, unless of course it had elected to settle on the far side of the moon. There seems no valid reason why aliens reaching the moon should restrict themselves merely to the hidden side. Moreover, space vehicles from Earth have over the last few years been orbiting the moon fairly regularly. Even if alien colonies there were not actually seen, it seems highly improbable that their inevitable radio traffic could have gone undetected.

It seems much more rational to assume that at present alien colonies do not exist either on the moon or anywhere else in close proximity to Earth. This does not rule out those parts of the solar system more remote from us. Neither does it mean that aliens could not have come close to our planet in the past nor that they might not do so in the future.

The fiction writer has not infrequently set his bounds far wider than the nearer stars. Indeed, he has roamed the galaxy at will from hub to rim and back again, always with consummate ease. From time to time he has even gone beyond it. What, then, can be said of such farther extrapolation of the space theme—of journeys between galaxies, of intergalactic travel? The nearest other galaxies to us are the Magellanic Clouds. These can be seen only from the southern hemisphere and a very lovely sight they are. Their distance from us is an almost unbelievable 150,000 light-years and in order to cross this black and starless abyss a fusion-powered starship would require something like 2 million years. Extensions of conventional techniques are clearly out—for advanced aliens just as much as for ourselves. This is not even the type of gap that advancing colonization might bridge. If, however, some of the ideas and concepts we will be considering in later pages can be given practical expression, then we may indeed envisage the possibilities not just of creatures from other stars, but of beings from *other galaxies*! Systems of transit which virtually annihilate distance must, it seems, be as applicable to intergalactic regions as to interstellar ones. Nevertheless, since in these pages speculation has of necessity been given considerable free rein, it would perhaps be advisable to avoid the extremest of extremes. For that reason we will restrict our thoughts to our own galaxy (the Milky Way). In this vastness there is surely scope enough.

Briefly, then, where do we stand at present on the subject of aliens?

1. We can dismiss the idea that we are unique and alone since there seems no good reason why we should be. Thoughts to the contrary carry undertones of megalomania.
2. We might believe we are so primitive, unpromising, and bellicose that aliens want nothing to do with us. This could be nearer to the truth. But just as we would be interested in the type of culture developed by an unknown backward tribe, so presumably would a sophisticated galactic people be interested in a backward galactic tribe.
3. Other beings may simply be watching us, regarding us much as a biologist would a low form of life. And just as the low order of life could not comprehend the biologist, neither might we the

aliens. If contemporary UFOs are genuine it would go far toward explaining their peculiar antics.

4. We could dismiss the whole subject as the greatest of non-events by declaring quite simply that interstellar travel is, and will forever remain, impossible for technical reasons. Acceptance of such a doctrine would represent an appalling form of defeatism totally unworthy of our race and society. Because something is presently beyond our capabilities does not mean that it must always be so. Even less does it mean that it is impossible today to beings vastly superior in intellect and mental capacity.
5. We could assume that aliens do exist, interstellar capabilities being possessed by some. Because of the vastness of our galaxy and the immense number of stars it contains, the chances of any reaching our solar system are so remote as to be negligible. Whereas this may appear to be the safest and most valid premise, it does not allow for a tide of spreading colonization originating in solar systems much older and maturer than our own.
6. We could ponder on whether the first "spray" from such an advancing tide is already reaching our planet (see Appendix). This seems well worth investigating. It is also worth remembering each time we look into the depths of the star-powdered sky!

Recently a science fiction writer posed this question: "Will we reach the stars before the stars reach us?" Perhaps that question has already been answered. Certainly in these pages we will be unable to come up with a categorical verdict on the issue of past and present alien visitations to our world. But we hope the possibilities will at least make the reader ponder, and continue to ponder! It may yet bode ill for us if we continue to think only in terms of our (largely self-inflicted) terrestrial problems. We look inward and see many perils. Occasionally we should look out—and *up!*

# 3. SPACE THAT BENDS

In the preceding chapters we mentioned several methods of interstellar travel which, to all intents and purposes, can be regarded only as extrapolations of our present space technology. So far as *swift* transit between stars is concerned these could have no relevance whatsoever and it is very improbable that alien races millennia ahead of us would settle for techniques so crude and impractical.

If such beings have visited our planet freely in the past, or are presently doing so with consummate stealth and efficiency, they must have adopted and perfected techniques that are entirely novel to us, the amazing fruits of a yet undreamed-of technology. We know well enough the parameters of difficulty, distance and time, the second leading directly from the first. The gulf between star and star is vast and because the velocities presently available to us are low, transit times in terms of the human life-span are not just disproportionately long, they are impossible. That quite simply is the crux of the problem. Interstellar travel is *not* impossible. In a sense, and in the light of what has already been achieved, it is perfectly feasible. Earth to moon, Earth to star—the basic problems are the same. All, that is, except for one, and that one is presently insurmountable: time. Our lives are too short and the way is too long. Power sources, both chemical and physical, just cannot give us velocities of the order required. Even if the velocity of light (186,000 miles per second) could be attained, tremendous and far from orthodox problems would certainly remain. We cannot go sufficiently fast. The only possible alternative is to telescope distance, somehow make the way shorter.

Right away this sounds like the beginning of an excursion into the realms of science fiction where virtually nothing is impossible.

But this is not science fiction; this is reality. Such a scheme, therefore, cannot be possible—or can it?

In this and succeeding chapters we will endeavor to examine certain possibilities—possibilities which might in the fullness of time give us the key to the stars. The expression "fullness of time" must be taken in its most literal sense. We have certain ideas on the subject. These may be exceedingly hazy. In fact they probably are. They could in the end lead us along the right path to vistas of the most exciting kind, or to the deadest of dead ends. However, in the context of this volume what we will be trying to consider are not methods enabling *us* to travel swiftly and easily from star to star or galaxy to galaxy. Even were we able to comprehend the underlying principles it could well be centuries before we were in a position to start translating these into practical reality. Rather, our intention is to explore several promising though highly unorthodox schemes, endeavoring to ascertain whether it is by such methods representatives of advanced galactic communities might visit, or indeed have already visited our planet.

If any of the crop of "flying saucer" tales are really authentic, then such visitors can have originated only on the planets of other stars. (It seems entirely reasonable by now to assume that our sister worlds about the sun are totally devoid of all higher life-forms.) If these are truly visitors from the stars, then their modes of transit must be highly sophisticated in the extreme. Of this there can be little doubt. These must be modes devoid of the pathos, poignancies, and difficulties which space "arks" or ships full of hibernating astronauts automatically invoke.

We will seek to show in the second part of this book that such visits may have been made in the past as well as during the present, right back through recorded history and into the geological epochs. We do not say categorically that they *have* been made, only that they *may* have been.

Most of us have a fairly fixed idea of the nature of the universe. It has certain features and facets which go to constitute its form. We see these. They are obvious and tangible. Our senses accept them. There is no reason why they should not. If told that perhaps all is not as it seems, we are naturally prone to skepticism. We point to all the clear visual evidence. Not only is this abundant; it is acceptable to our minds. It is normal, rational, and (so we think) logical. But is the universe really as we see it? In a sense we should now perhaps think about the universe in the same way that the seventeenth-century philosopher Descartes thought about the physical world. He maintained that though the existence of the physical world could be proved by necessary arguments there was no corresponding necessity that it be in any way similar to the world the senses depicted.

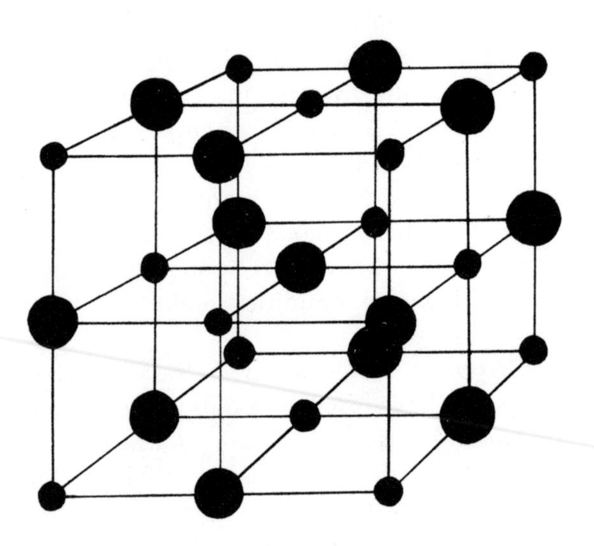

Figure 1

*Arrangement of chlorine and sodium atoms in common salt—a solid, yet largely empty space.*

Suppose a large cube of metal were placed in front of us. Suppose also that we are not in the twentieth century but back in the fifteenth. Sitting with that very solid and very real metal cube before us in the year 1477 we are told that the cube is *largely empty space.* Ridiculous, of course! The thing is visible, tangible in the fullest sense, hard, heavy, solid, and unyielding. Were it largely empty space, surely we could push a finger into it? Made today, however, such a statement is neither stupid nor ridiculous, merely one of established scientific fact. Modern concepts of atomic and crystal structure have taught us the true state of matter (Figure 1). Might there not then be a parallel, a possibility, that our nice, conventional, so rational universe is *not* as it seems? Do we really comprehend its form? There exists already a fast-growing suspicion that in fact we do not. Perhaps the physical universe is quite different from that which sight and our other senses would seem to indicate.

Let us return to our metal block (Figure 2). The shortest route between points A and B is obviously *through* the block. Clearly that route cannot be followed. Instead we must make a detour around the side, a longer path and one involving the expenditure of more time. Yet why cannot we go directly from A to B through the block? Modern physics tells us the block is largely empty space. If so, there should be no impediment to our passage. But there is, quite unmistakably, a very definite impediment (and how surprised we would be were there not!). Here is empty space, yet we cannot pass through it. Admittedly that empty space is not of the obvious kind—quite the reverse in fact. Nevertheless, we now know and accept that it *is* there. We must also accept that we cannot pass through it. A peculiar sort of impasse!

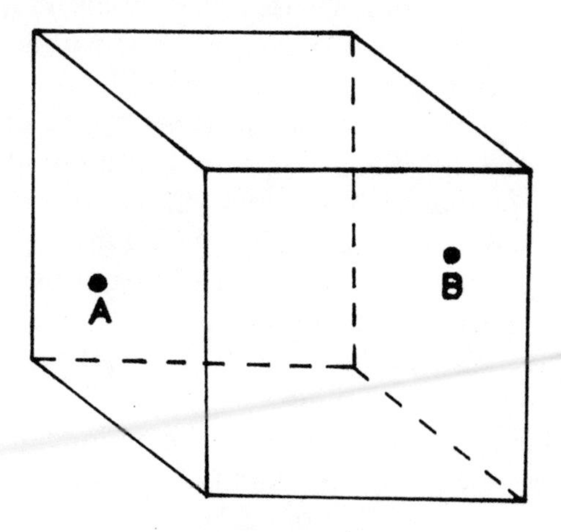


Figure 2

Is there a parallel, then, with the universe? We wish to go from star A to star B. Suppose there is a shorter route than the seemingly straight-line path from A to B. We do not see it, but should that be taken to mean it does not exist? In the metal block we do not "see" the direct route from A to B, yet undoubtedly it exists, although only specialized techniques will permit particles to traverse it.

Here may be the kernel of the whole business—a specialized technique. But specialized techniques are possible only when we understand the true nature of the medium. What we must now ask ourselves is whether or not we *really* understand the medium we call space. As we have said, there is already a growing suspicion that we do not, and until we do we cannot think of specialized techniques capable of freeing us from the apparent inhibitions and restrictions which space imposes.

Let us, without letting our ideas range too far or too outrageously, look just a little more closely at so-called conventional space. What, first of all, are our normal, "background" thoughts concerning it? Probably, and certainly not incorrectly, that it contains an immeasurable number of stars grouped into galaxies; that the number of these galaxies is as immeasurable as the stars themselves since each increase in the power and range of telescopes (both optical and radio) reveals more and more of them; and that the galaxies themselves are receding, those most remote retiring with the greatest velocities. There, generally speaking, is where our thoughts end. What follows is regarded, not without good reason, as the unfathomable. Despite the obvious difficulties let us see if we can go a little farther. Is the universe finite, does it have a boundary, an ending? If we accept the premise that it is finite we must also accept that of some other entity lying beyond. What, then, is the nature of that "beyond"? And if it too should be finite, what then? What lies beyond that? The idea of the universe being finite is to us logical. Indeed, it is almost essential to our supposedly rational and certainly finite minds. Unfortunately, as we have just seen, this concept solves nothing.

What of the alternative—that space is infinite, something that goes on forever, unending space and time? The human mind tends to reject such a concept. It cannot be grasped—no beginning, no end, just a form of "foreverness." This is not a new problem. It was occupying the minds of men as far back as the beginning of the seventeenth century. In his *Principles of Philosophy* written in 1644 we find philosopher René Descartes already grappling with it. "I do not say," he wrote, "that the universe is *infinite* but only *indefinite* and in this there is a notable difference. For to say a thing is infinite one must have some reason which makes this known, which can only be the case of God alone; but to say it is indefinite we only need to have no reason for proving that there are limits. So it seems to me we cannot prove nor even conceive there are limits to the matter constituting the universe."

Either way, then, we are apparently baffled. How in the circumstances can we find an answer? Visual observation provides no solution. There is a fairly definite limit to the amount of universe we will ever see for the simple reason that galaxies beyond 10,000 million light-years are hurtling outward and away from us at almost the speed of light. Radiations from them, if they reach us at all, are too weak to register. In other words, there exists for us a *cosmic horizon.* Use of the term "horizon" is particularly apt, as we shall shortly see. Our observable universe must therefore be regarded as a "sphere" having a diameter of about 20,000 million light-years. This represents a lot of space and a lot of contained matter, but it is certainly *not* the entire universe.

Arguments on whether the universe is finite or infinite can never be resolved so long as we think purely in terms of conventional geometry. It may well be that space displays curvature—in other words, it "bends over" on itself in some odd manner. To talk of space, which we regard simply as a void, as "nothingness," doing something that endows it with the properties of an entity may seem very odd. Perhaps if we consider a few simple analogies the picture will become clearer.

Suppose we desire to journey from point A to point B on the surface of the Earth, say, for example, a distance of one mile along a perfectly straight and level road. This we know from elementary-school geometry to be the shortest possible path ("The shortest distance between two points is a straight line"). Suppose now that points A and B lie on opposite shores of the Pacific Ocean and that the ship carrying us steers an absolutely straight course between them. Have we taken the shortest possible path? We may have taken the *shortest path possible to us* but it is certainly *not* the shortest path geometrically, as a little thought will make plain. Our ship has in fact all the time been steaming along a path that follows the curvature of the Earth's surface. This path is an arc of a circle, not a straight line, and even Euclid himself could not equate the two! Indeed, our earlier journey of one mile along a straight road must be highly suspect for precisely the same reason. The cities San Francisco and Los Angeles are 350 miles apart (Figure 3). An

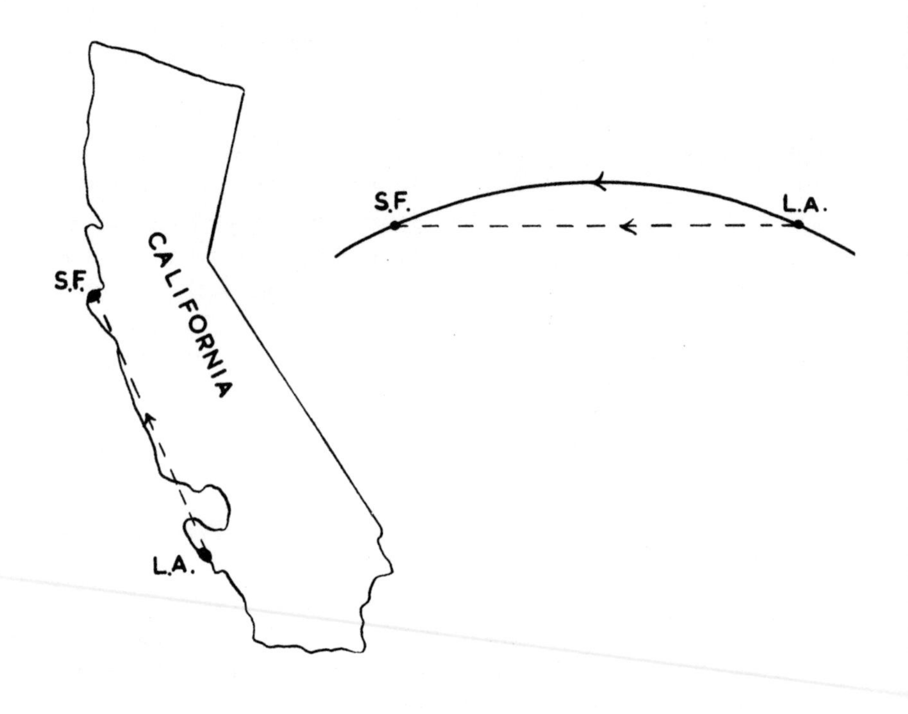


Figure 3

aircraft flying in a straight line from one to the other would in fact be flying in a circle that follows the curvature of the Earth.

We can illustrate the validity of this in other ways. Take, for instance, a triangle drawn on a perfectly flat sheet of paper (Figure 4). It must be obvious even to a completely nonmathematical person that however much we care to extend lines AC and BC they cannot in normal circumstances ever *meet.* It must be added, though, that this is valid only with respect to a *truly plane* surface such as our sheet of paper. If we draw our triangle on the surface of a sphere a totally different state of affairs results. In this instance extending the two sides of our triangle, though it leads initially to their divergence, will ultimately result in their *meeting* on the *far* side of the sphere (Figure 5).

To sum up, we can reasonably think of our planet's surface as flat only if we are concerned with a very small portion of it. When we come to deal with larger areas the concept of straight lines

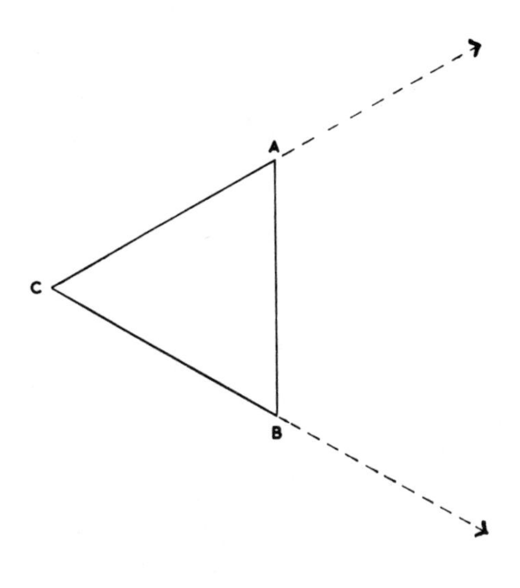


Figure 4

must, of necessity, be abandoned in favor of curved ones. In other words, since our planet is a sphere (or more correctly an oblate spheroid) we must think in terms of curves.

There could be an analogy here with the universe at large, for contemporary cosmology now accepts that gravity has the effect of bestowing curvature to space. Unfortunately, this is a difficult concept to envisage used as we are to a three-dimensional world of length, breadth, and height. And because we are three-dimensional creatures ourselves we have no particular difficulty in thinking about *two*-dimensional curved space such as the surface of the Earth. *Three*-dimensional curved space is quite another matter. It is just *this* space that may comprise the entire fabric of our universe. The fact that it is curved *empty* space does not exactly help matters!

This understandable difficulty is appreciated by even the most eminent scientists. In his *Mystery of the Expanding Universe,* William Bonnor writes,[1] "I do not ask you to visualize three dimensional curved space—I do not believe anybody can do that, not even the most hard-bitten geometer. I ask you simply to admit it as a possibility." In similar vein Richard P. Feynman in his *Feynman Lectures on Physics* says,[2] "We live in three dimensional space and we

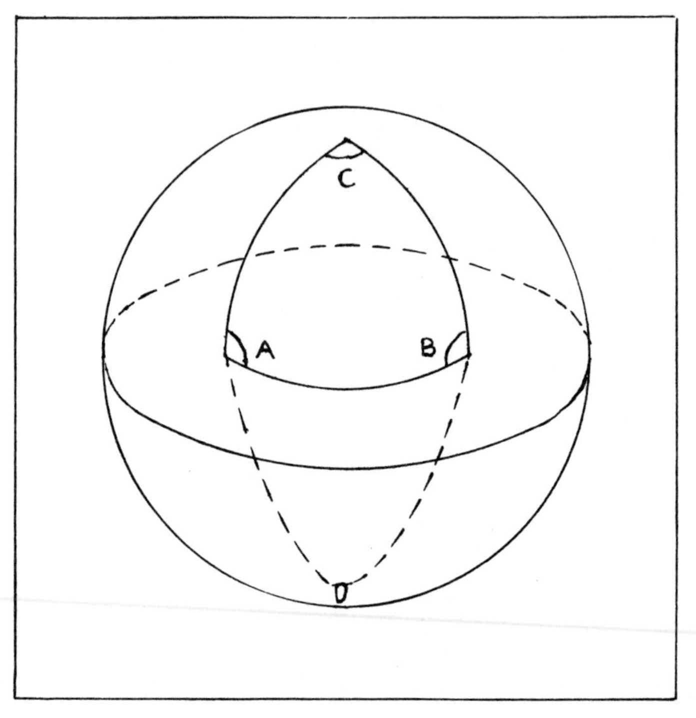


Figure 5

are going to consider the idea that three dimensional space is curved. You say, 'But how can you imagine it being bent in any direction?' Well, we can't imagine space being bent in any direction because our imagination isn't good enough."

What both of these writers are saying is that this is something we should just accept because its existence is highly probable. We should not be overconcerned at our inability to visualize it. Nevertheless, the concept of three-dimensional curved space is sufficiently important to warrant some attempt at its understanding. Only in that way can we hope to gain an insight into the universe on a cosmological scale.

Let us imagine for the moment that the entire universe is a vast sphere made from some kind of transparent plastic.[3] Its skin is so thin that we can regard it simply as a two-dimensional curved surface. Embedded in this surface is our Earth (within the Milky Way galaxy) and all the other galaxies (Figure 6). The parallel with movement on Earth itself now becomes apparent. Just as a journey between Los Angeles and San Francisco really involves a curved path, so does a voyage between galaxies A and B. On a smaller scale, but still following a curved path, is a journey between two stars in the same galaxy.

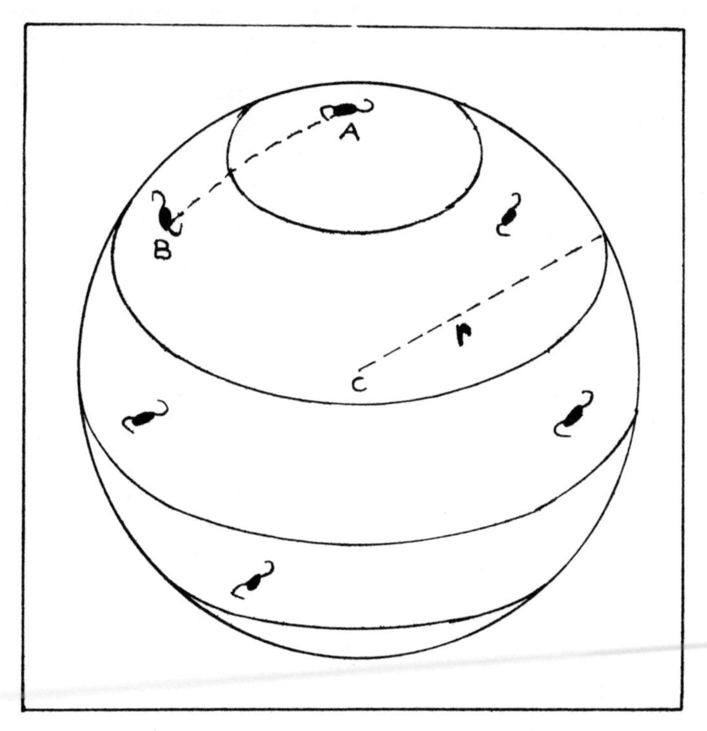


Figure 6

Now there seems little doubt that a journey from Los Angeles to San Francisco (Figure 3) would be shorter were we able to follow the lower, dotted path through the Earth. (The fact that it would also be infinitely harder is a sufficiently good reason for sticking to the freeways!) In like fashion a journey from galaxy A to galaxy B would be shorter were we able to "cut" through the sphere of space. This too, for us at least, is presently impossible. We can regard such a path as lying within *another* dimension which we cannot enter and to which our senses do not respond.

As we said earlier, it is one thing to postulate the existence of such a dimension. It is quite another to try to visualize it. However, it is not really essential that we possess a precise mental picture of the situation. The day may come when, as a result of further research and inquiry, this will be feasible.

At this point it is desirable that we abandon the analogy of a sphere since there is no reason to believe that space curvature is of the spherical kind. It is in fact likely to have a very different and much more complex form, so complicated that it can be portrayed only mathematically. However, a generalized and very rough idea of the probable geometry of curved space is shown in Figure 7. It cannot be overemphasized that at best this represents only a crude caricature. Nevertheless, it may serve to give us a working mental image, albeit imperfect, of the very peculiar geometry of curved space.

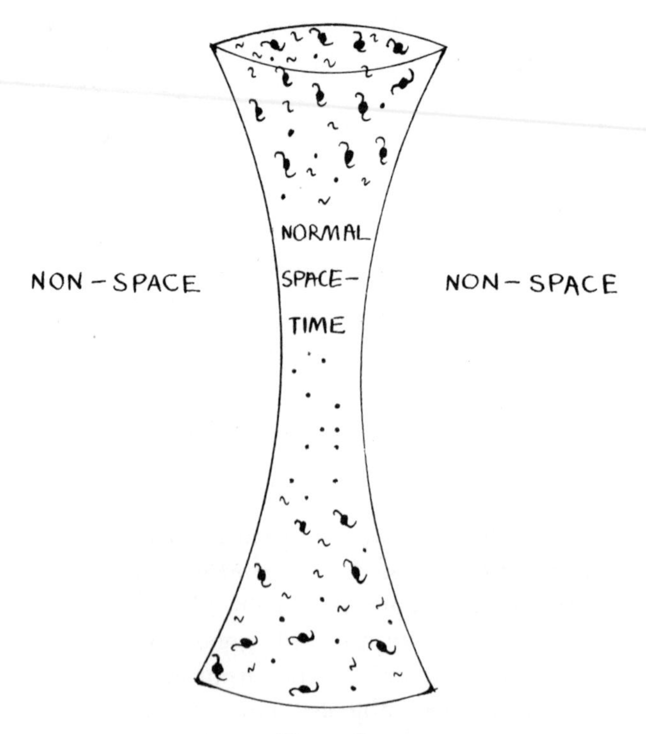


Figure 7

*Probable "saddle-shaped" form of the universe. Note that the galaxies do not fill it, but merely traverse the normal space-time "skin" of the surface.*

Bodies moving in curved paths in curved space would theoretically be able to proceed indefinitely, just as a ship could continually sail around the world. A galaxy would no more run out of space than the ship would run out of ocean. In other words, the path of a galaxy will never lead to the "edge" or "end" of space. The most distant of the observable galaxies may represent the "rim" of our local region of space just as a ship disappearing over the horizon indicates the "rim" of our local area of ocean. But what we see is no more space in its entirety than is the ocean we can see up to the horizon.

Incredibly vast though space may be it is still regarded as finite in amount. At this point let us consider a fly crawling around the surface of a ball. The ball is quite definitely finite but in a sense its surface is not. Assuming the fly has the stamina, it can crawl around the ball indefinitely. It may run out of energy but it will certainly never run out of surface. And so, in like fashion, the galaxies continue to rush onward and outward, never running out of space. Of course if we continue this analogy we would expect them to go right around space and wind up back where they started. This is highly unlikely since the universe is seen not as a steadily expanding balloon of indeterminate shape, but as a balloon which, over aeons of time, is alternately expanding and contracting. This represents the now widely accepted concept of a pulsating universe, that is, a sequence of "big bangs" rather than as a single once-and-for-all-time super "big bang."

We have drawn attention to the fact that the universe may be saddle-shaped (Figure 7). The precise shape is difficult, if not impossible, to imagine in the light of our present very imperfect knowledge. It is probably best to liken it to a large soap bubble or balloon of indeterminate shape. The really important point to grasp is that our universe on this model is *not* the bubble in its entirety but merely the *skin* of that bubble. This peculiar cosmic skin is not, however, a curved *two*-dimensional surface. It has thickness, too. (Even the skin of a soap bubble must have a certain thickness and if we were mere atoms it would seem very immense indeed!) It is, in fact, a *three*-dimensional entity and represents our conventional universe. In this three-dimensional "skin" swim the galaxies (Figure 8). The medium in which the galaxies (and their constituent stars) exist is, therefore, infinite inasmuch as there is no

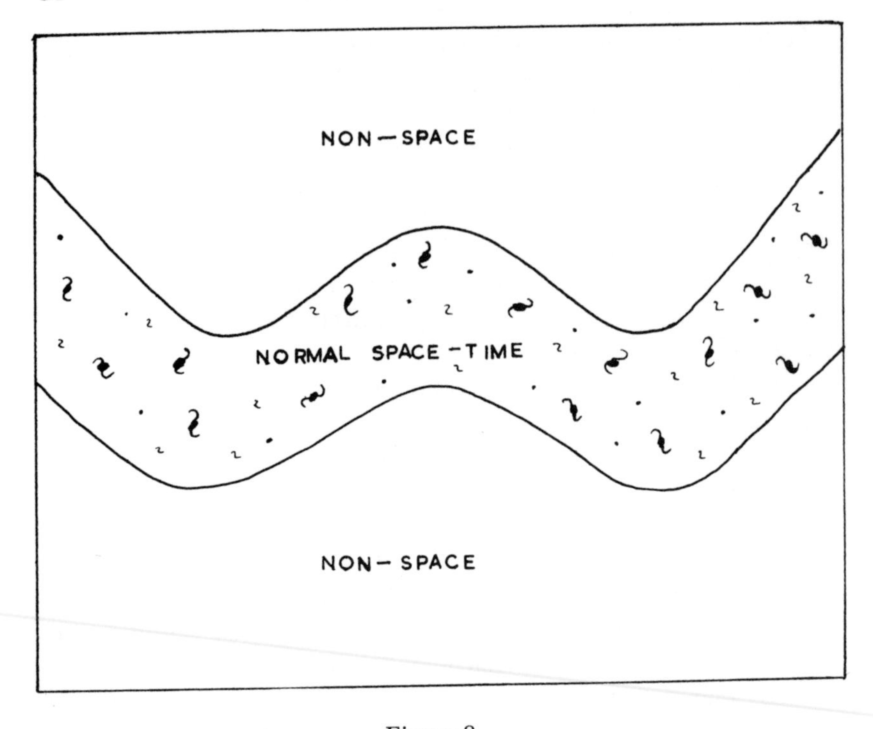


Figure 8
*The normal space-time universe as a four-dimensional "skin" surrounding and surrounded by "non-space."*

end to it. (If this seems perplexing consider the fact that there is no end to a circle.) It is also finite in the sense that a bubble is a finite object. What lies within (and without) our mysterious universe we do not know. It is to us simply an unknown dimension which is generally termed "non-space" or "hyper-space." The vagueness of the terms indicates the vagueness of our knowledge. The renowned mathematician Bernard Riemann has postulated not just a single other dimension but a whole series of them. Might this have relevance when we ponder what might lie beyond non-space? It may indeed, though it seems unlikely ever to solve anything in the ultimate sense, for these are issues which may eternally lie beyond human comprehension.

We thus see non-space as the unknown fifth dimension (the known four being length, breadth, depth, and *time),* a peculiar and mysterious entity within and without the cosmic "bubble skin" which on our model is the presently known universe. Lying within this four-dimensional "skin" are the galaxies. Figure 9 shows two of these (labeled X and Y). Once again we see the conventional curved route which, were we to embark on a space voyage from our Milky Way to the Great Galaxy in Andromeda (our nearest galactic neighbor), would seem as a straight path (and a highly tedious one since the two galaxies are about 2¼ *million* light-years apart). But by cutting through into non-space, into that other dimension, a shorter path becomes available. The reader may

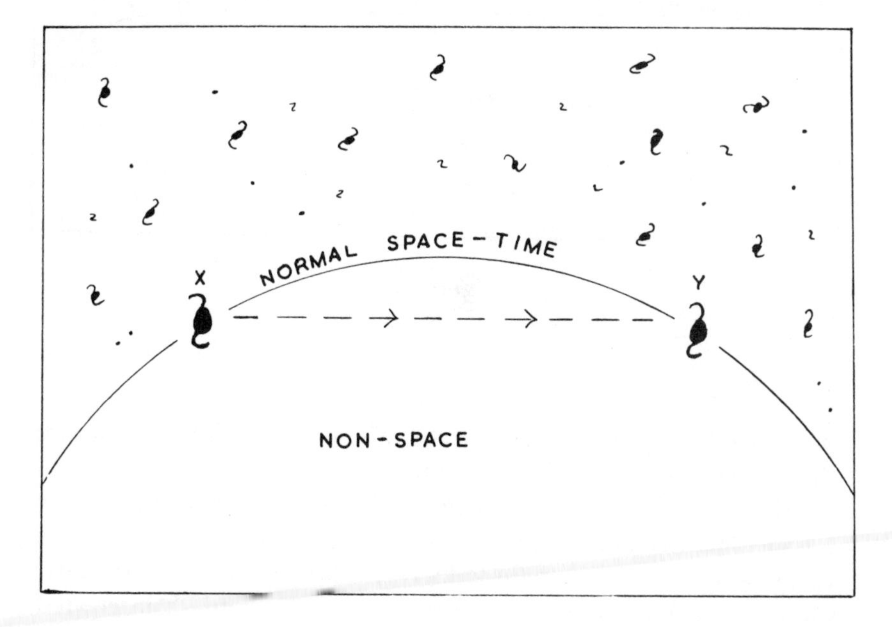


Figure 9

point out (and very rightly) that the latter, though unquestionably shorter, still represents a very considerable distance. What, of course, we do not know is how in such a dimension we would stand in relation to time. Non-space, for all we know, could be a dimension in which, quite literally, time stands still, thus permitting almost instantaneous transit through it.

An interesting point arising from this new possibility seems worthy of mention. In 1915, George Cantor, professor of mathematics at the University of Halle in Wittenberg, Germany,

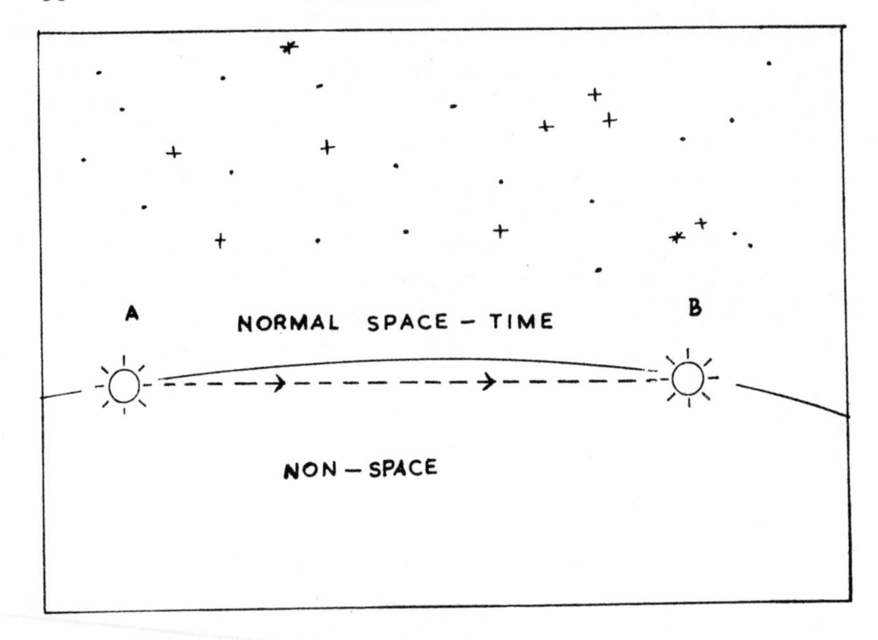


Figure 10

published his unique theory of transfinite numbers. This postulates the existence of values *beyond* infinity—a most peculiar concept to say the least. There are two distinct possibilities: Either the theory is a mere paradox or it represents mathematics pointing the way to vistas of a new and incredible kind. At first, hardly surprisingly, Cantor's work evoked little more than smiles and disinterested amusement. More recently, however, further work by Subrahmanyan Chandrasekhar has indicated there might be more to Cantor's strange theory than meets the eye. Another mathematician, J. C. Campbell, has further enlarged on Cantor's work to show that, *given sufficient power,* it should be possible (theoretically at least) to travel between two points *without actually traversing the intervening distance*! The power requirements, we must hasten to add, are formidable in the extreme—the equivalent, it has been estimated, of over a million thermonuclear devices of high yield! Such a claim seems utterly absurd and a complete affront to logic and reason. Tie it in, however, with the idea of a non-space bypass and where do we stand? Might this not be the link with possible instant transit through the fifth dimension? If we proceed from galaxy X to galaxy Y in this manner and by such a route we have certainly *not* traversed the *conventional* intervening distance between X and Y.

The individual stars in each galaxy are also moving in the four-dimensional space of our cosmic bubble's skin (Figure 10) though of course the arc between them represents a very much smaller "section" of that skin. Seen in this light we have thus also a non-space path between stars in a galaxy.

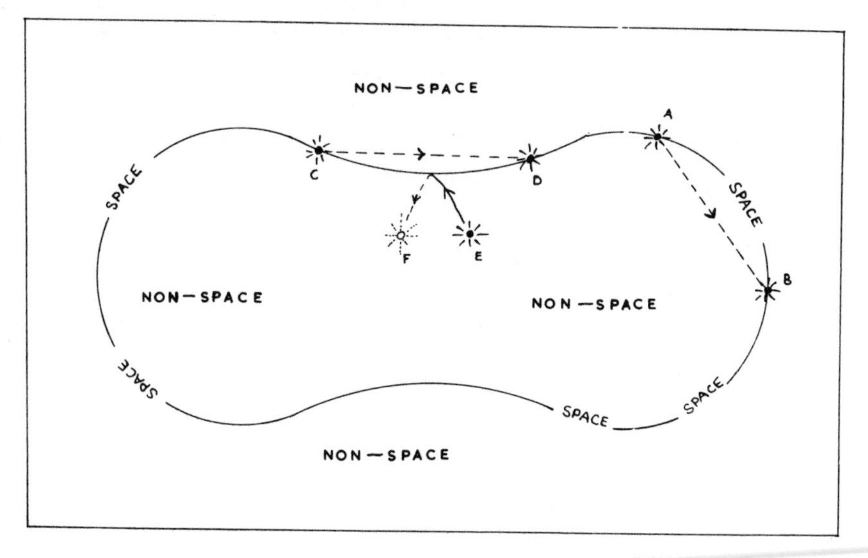


Figure 11

This "cut" through into the bubble itself is all right, or so it seems, so long as we consider *positive* curvature. But if the form of the universe is the odd saddle-shaped thing which many cosmologists and mathematicians now believe it to be, then relative to its "outside" are regions of *negative* curvature. Between stars A and B we have positive curvature (Figure 11), between C and D negative curvature, and "across the saddle," E to F, more positive curvature.

We can therefore proceed as before with respect to journeys between A and B and between E and F. But how about a journey from C to D? Clearly, there is no straight-line path *within* the space bubble here. The position may, theoretically at least, be explained by regarding the region outside the space bubble as non-space also.

The shift toward the red end of the spectrum of light emanating from the galaxies is generally regarded as proof of the fact that the galaxies are receding at fantastic velocities. The universe is seen therefore as an expanding one. In recent years a new model has been postulated, that of a *pulsating* universe. The present expansion is seen as a single expansive cycle after which a contractive cycle will follow, leading in turn to yet another "big bang" and a further expansive cycle, and so on.

During the present cycle the skin of the cosmic bubble—that is, our supposedly familiar universe—is stretching much as would the soap film of a true bubble or the elastic of a balloon. In other words, it is expanding and the galaxies within it are drawing apart.

Techniques enabling starships or instrumented probes to pass into and through regions of non-space are clearly outside our abilities both now and throughout the foreseeable future. There are many questions that could be asked, among them the following:

1. What force or agency would enable a vessel to pierce the skin and enter non-space and, of equal importance, regain the conventional universe?
2. To what extent would time dilatation affect vessels in non-space?
3. Could permanent or fatal damage to living organisms result from passing through non-space?
4. How in the "nothingness" of non-space could spaceships be guided to their respective destinations? This would seem to represent the ultimate in navigational problems.

So far as *we* are concerned there can as yet be no possible answers to questions such as these. As a wild exercise in speculation we might ponder the idea of a force field of some kind being used to pucker the skin of space preparatory to passing through an induced rent which at once "heals" behind the ship. In view of what we said earlier regarding a skin of normal space-time surrounding the universe it might be supposed that to achieve this a space vehicle would require to lie in a region of space-time *very close* to the boundary with non-space. This, of course, pinpoints precisely the weakness of any model we produce to explain the form of the universe. Such a model may give a broad outline but in fact it is virtually impossible to indicate in a four-dimensional way what is a five- or multi-dimensional system. According to Einstein it is mass that produces space curvature. The greater the mass the greater the degree of space curvature. Thus even a body as trivial and unimportant as a mere star produces its own small contribution in this respect. Presumably it is the summation of all this which produces the degree of space curvature brought about by a galaxy. And in like manner, but on an infinitely greater scale, the total effect of all the galaxies is to bring about that degree of space curvature which bestows on the universe its peculiar form and geometry. It has been suggested by some cosmologists that the normal space-time with which we are familiar is honeycombed by a series of non-space "tunnels" due to a linking up of cylinders of non-space produced by individual stars and galaxies. This, of course, must remain a highly speculative idea, something once again in the nature of a model to help render tangible the completely intangible.

With respect to such matters we can at best only probe, and that very feebly! What we see, we see only hazily. Many there are who would question even that, saying that, so far, we just do not see at all. Yet we can and must continue to reason and postulate. No one has actually *seen* an atom, yet we freely accept the existence of such an entity. Today we postulate the fifth dimension, tomorrow we may accept it. As with the atom, we do not require to have seen it.

To highly advanced civilizations in this and other galaxies the picture could, by now, be very clear indeed. It might also be one they have used to decided practical advantage so far as interstellar transit is concerned.

If an unknown dimension exists, what would it be like to the occupants of a space vessel which has succeeded in entering it? This we cannot tell and it is almost impossible to imagine. There are, no doubt, a variety of ideas, some perhaps irreconcilable. The obvious choice to the writer is blackness, utter and complete, of the most fundamental and terrifying kind, a sheer "nothingness" of nightmare proportions. If one of our kind ever enters it he may well at that moment recall the epic, terrible words of the famous and doomed British polar explorer Captain Robert F. Scott, who as he surveyed the desolate wastes around the South Pole in January 1912 was heard to say, "My God, this is an awful place!"

In our universe such a place there may be and through it, even at this very moment, may be passing aliens in interstellar and intergalactic space transports. It is an odd thought, as intriguing as it is frightening, for one day the skin of space near our familiar sun may pucker and part to allow the entry of a craft beyond our wildest imaginings. We should, perhaps, even ponder how often in the past this has already occurred!

# 4. TUNNELS IN SPACE

For a long time now it has been the practice of science fiction writers to adopt certain convenient subterfuges when confronted with the time-distance barrier of interstellar space. A favorite has always been the "space warp," so much so that by now the term is woven into the very fabric of science fiction. It is rare, of course, for an author ever to dwell too closely on the matter. Indeed, an aura of ineffable vagueness generally surrounds the term. In this respect it joins the almost equally legendary "hyper-drive," usually a device whereby velocities in excess of that of light may be achieved. The space warp is a peculiar facet of space enabling a spacecraft to pass from A to B without having to traverse the intervening distance. In the preceding chapter we saw how this might have some relevance. It is doubtful, though, if the space warps considered there can be equated with the space warps of science fiction. These are regions of curved (or warped) space in the region of a star or galaxy brought about by the mass of the particular star or galaxy. The space warp of the fiction writer is intended as something infinitely greater, a kind of cosmic vortex capable of leading, it would appear, almost anywhere. Such trivial problems as navigation are conveniently forgotten.

Is there anything in the universe even remotely like this? Until quite recently the answer would almost certainly have had to be an emphatic negative. We still cannot say there is tangible and unequivocal evidence to support the affirmative point of view, certainly not for anything along the lines of the more lurid type of science fiction. We can, however, now postulate the probable existence of a cosmic phenomenon having very odd properties indeed. If, by now, the reader is wondering where this long preamble is leading the answer can be summed up in two short words: "Black Holes."

Recently we have been hearing a lot about Black Holes. Already a certain aura seems to surround them. As a term, it is more descriptive than scientific, which is probably why the more specific term "singularity" is preferred. (As we will see, the terms are not really synonymous.) Even this carries an air of deep mystery. What, in fact, are Black Holes? Do they exist—and, if so, why?

If the reader wishes to peruse the first reference to these peculiar objects in the published literature he or she must go back to 1939. In that year they were predicted by J. Robert Oppenheimer. Not until 1970, however, did anything come along which could be construed as real evidence. This was derived from data gathered by the "Uhuru" Small Astronomy Satellite from which it appeared that the X-ray source, Cygnus X-1, might represent a Black Hole/class B supergiant binary system.

Before we begin to consider the possibilities and potential of such objects as super "space tunnels" leading to the "great plains" of the universe it is highly desirable that we examine the simple basic theory and available evidence (such as it is).

In something like 5 billion years our friendly and familiar sun will have consumed so much of its hydrogen in thermonuclear reactions that a dangerous and highly critical state will have been reached. The result, according to current theory, is that it will swell out into a vast bloated red star of the type known as a "red giant." The present diameter of the sun is approximately 850,000 miles and it is anticipated that during its eventual transition to a red giant there will be a 250-fold increase in its dimensions, giving it a diameter of around 200 million miles. In these circumstances there can be little doubt as to the fate of the inner planets. Mercury and Venus will certainly be consumed. Earth, even if it escapes (highly doubtful!), will be transformed into a cinder. The safety and sanctity of Mars will be in jeopardy, while the effect of heat and radiation on Jupiter may bring distinct changes to that huge planet, assuming that after such a protracted period of time great evolutionary changes have not already occurred there as part of the natural order of things. This "new" sun of ours will be vastly different from the one with which we are so familiar and the difference will not be restricted to color and dimensions. It will also be very much less dense. Solar density is presently about one-fifth that of Earth, but when the sun becomes a red giant it will probably be only about one-tenth that of *air.*

As more and more nuclear fuel is consumed (and this includes helium and certain of the heavier elements as well as hydrogen) the expansion process will cease and then go into reverse. Contraction will continue not just until the sun attains its former dimensions but until it has a diameter little greater than that of *Earth.* From tenuous red giant it will have shrunk to dense "white dwarf." Such an immense amount of matter contained within a "shell" of relatively minute proportions means a vastly increased density. Indeed, the electrons of its constituent atoms will now be so tightly packed as to produce sufficient outward pressure to prevent further contraction. So tremendous is the density of a white dwarf star that it is virtually impossible to visualize. A tennis ball filled with its material would have the mass of a light naval cruiser!

Red giants and white dwarfs are by no means exceptional. Large numbers of each type are known (for example: red giants—Betelgeuse, Antares; white dwarf—Sirius B). Such bodies represent senile and dying stars respectively and must be regarded as normal stages in the evolution and history of perfectly normal stars. This does not mean that all stars follow this course. If a white dwarf happens to possess a mass over and above a certain critical figure it is unable to resist gravity and suffers further contraction as a consequence. This figure (it is generally referred to as the "maximum mass limit") is only slightly in excess of solar mass (about 1.5 times solar mass).

Let us consider a star having a mass greatly in excess of that of the sun. After consuming a high proportion of its hydrogen "fuel" it swells out into the red giant state and then commences to contract. In such an instance, however, contraction may *not* stop at the white dwarf stage. Indeed, in the circumstances it almost certainly will not and so tremendous will be the resultant density and temperature that processes of an ultimately destructive nature are initiated. The climax is an explosion of cataclysmal proportions, the star becoming a super-nova. Some idea of the nature and scope of such an event may be gauged by the fact that for a few days a super-nova may outshine an *entire galaxy.* Up to 90 percent of the star's mass may be ejected into space, leaving only a collapsed core at the center of a very rapidly expanding cloud of gas. A classic example of this kind of thing can be found in the constellation Taurus, the well-known Crab Nebula. Such a stellar core is much too small and compressed to form a white dwarf and finds equilibrium instead as a neutron star. If on the stellar scale a white dwarf is small, a neutron star is minute, having a diameter in the region of *twelve miles.* And if the density of a white dwarf is extreme, then that of a neutron star is quite simply fantastic since it is about a *hundred million* times as great. (Imagine a tennis ball containing a mass equal, not to a light cruiser as before, but to that of a minor planet such as Juno which, for the record, has a diameter of over a hundred miles.)

It was just such a star that was predicted by J. Robert Oppenheimer and others during the years 1938 and 1939. At the time there was considerable debate on the subject, many astronomers doubting whether such bizarre objects could really exist despite the theoretical case which had been advanced. This was virtually the position right up to 1967, when pulsars were first discovered. Research into the nature of these very peculiar objects has since shown that they radiate regular energy pulses both in the optical and in the radio portions of the spectrum and that they are almost certainly due to the presence of a rotating neutron star.

The first pulsar was discovered at the Mullard Radio Observatory, Cambridge University, England in 1967 by a young woman research student, Jocelyn Bell (now Dr. Jocelyn Bell Burnell). At first it was thought this might be a signal coming from out of the depths of space from a distant alien civilization, and the popular press of the time (not unexpectedly!) had a positive field day with banner headlines screaming of "little green men"!

Now a most interesting question arises. We are aware that stars exist having masses more than fifty times that of the sun. Is it possible that such stars could, prior to their collapse, shed sufficient material to *reduce* their respective masses, thereby enabling them to fall within the critical limits of white dwarf or neutron stars? Although this may seem improbable, even more improbable surely is matter condensing to an even greater extent than that in a neutron star. In theory, greater densities *could* be achieved but only at the expense of the star's equilibrium. In short, standard gravitational theory is inadequate in the circumstances and relativity theory must be invoked if an answer to the problem is to be found. That answer is incredibly bizarre, for it postulates, indeed requires, the existence of Black Holes.

What, then, is a Black Hole? In essence it is a region of space into which a very massive star has "fallen" and from which neither light nor matter in any form can escape. The star has actually been crushed into virtual nothingness by a gravity so intense as to defy imagination.[1]

It is interesting to note that as far back as 1798 the celebrated French astronomer and mathematician Pierre Simon de Laplace made the (then) astonishing prediction that a body sufficiently massive and dense would be *invisible* since the escape velocity at its surface would exceed that of light.

Thus far we have only *defined* a Black Hole. We have still made no attempt to describe one. The main feature is thought to be a spherical surface having a radius proportional to the mass of the whole. This surface is termed an "absolute event horizon." The body responsible for the Black Hole (that is, the "crushed" star) is inside the "event horizon."

So powerful is gravity within the event horizon that even light is clawed back irrespective of the direction in which it was first emitted. The closer the emission gets to the event horizon, however, the greater becomes the tendency for the wave front to be drawn back toward the center of the Black Hole. It is intriguing to dwell on what must happen to light on the event horizon itself. We must suppose that here it will radiate neither outward into space nor inward toward the center of the Black Hole but will instead remain immobile—"frozen light."[2] A pictorial interpretation of the position is outlined in Figure 12. Rays of light emitted by the surface of the collapsing star escape from the star only if they lie within the imaginary cone (generally referred to as the "cone of escape"). Those outside the cone cannot. Their photons stream upward merely to descend again. Those at the midpoint have the worst of both worlds since their photons can neither escape *nor* fall back. There is but one option remaining to them—to go into orbit about the star constituting thereby a "photon sphere."

But how about rays of light from an *external* source coming toward a Black Hole? Since space-time has been rendered highly curved by the intense gravitational field, it is apparent that such

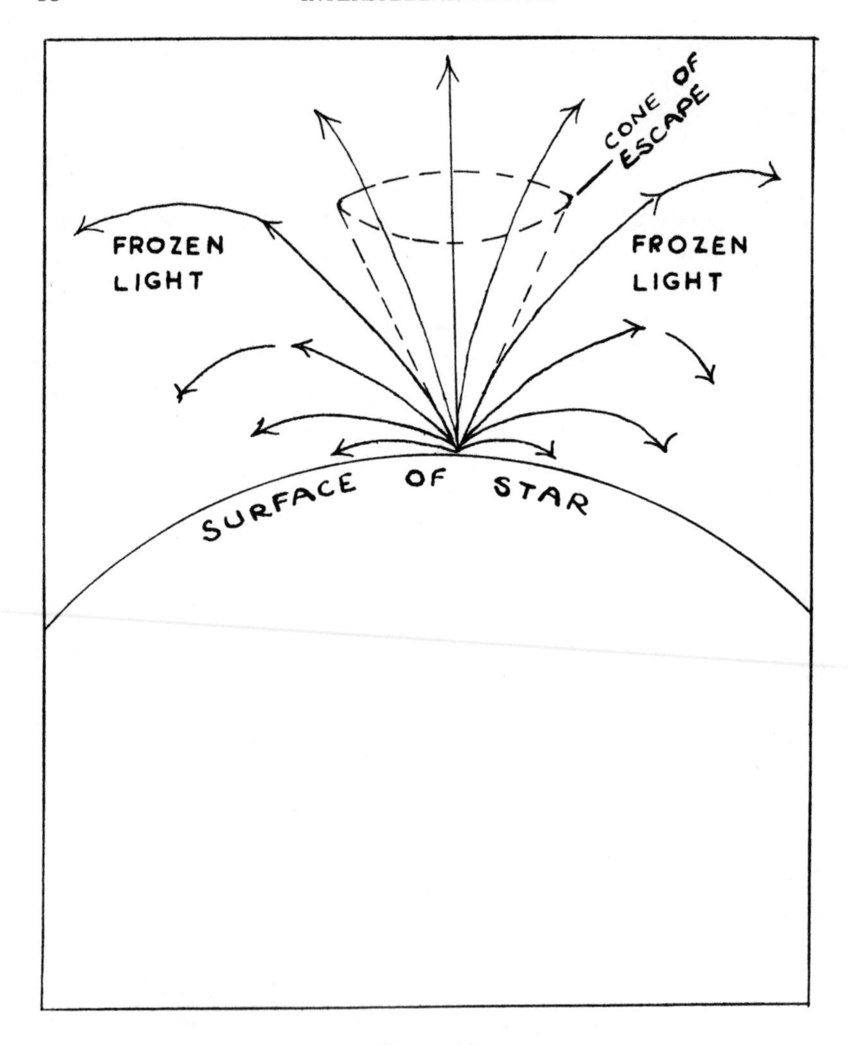


Figure 12
*Light rays leaving the collapsing star in paths* within *the "cone" will escape. Those emitted outside will not, being bent to such an extent that they* return *to the star.*

rays are going to suffer severe deflection. As the light rays are directed ever more closely toward the Black Hole there will be varying degrees of deflection. Those passing a considerable distance from it will suffer only slight deflection. Those passing more closely will experience severe deflection. There is a point when the rays are "captured." They neither escape (albeit severely deflected) nor are they "sucked" into the vortex. These rays are analogous to the rays emanating from the star which neither escape nor fall back and, as before, their photons also go into orbit around the star as part of the photon sphere. This orbit, it should be stressed, is very unstable. Or perhaps, more correctly, it is extremely precise—so much so that the slightest deviation by a photon from the true and highly critical orbit is sufficient either to send it off into space or spiraling downward into the dying star. Those rays coming directly toward the Black Hole are, of course, simply sucked into it (Figure 13).

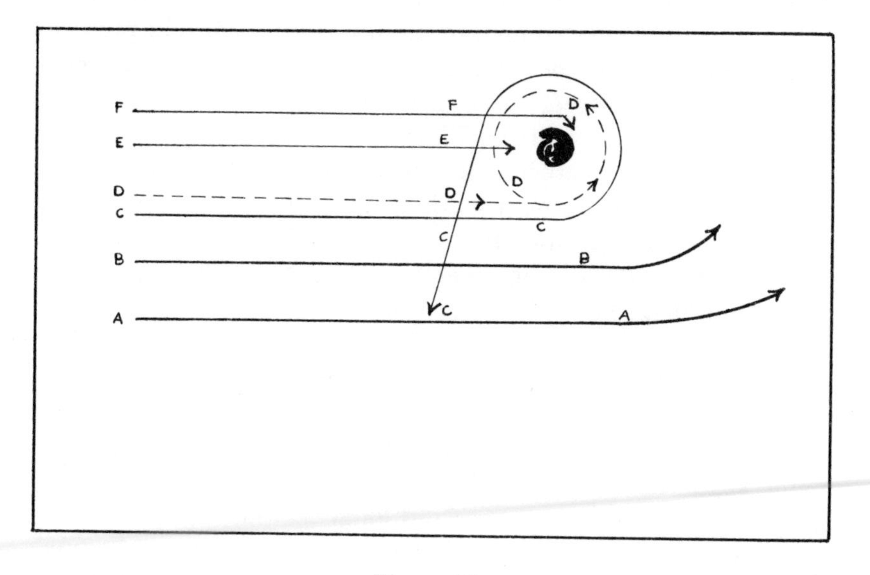


Figure 13

*EFFECT OF BLACK HOLE ON RAYS OF LIGHT TRANSMITTED TOWARD IT: (A) moderate deviation; (B) severe deviation; (C) ray turned back on itself; (D) photons of ray join "photon sphere"; (E) rays sucked directly into Black Hole; (F) rays drawn into Black Hole.*

But now back to rays emitted by the collapsing star itself. What happens as the star continues to collapse? The effect is to warp space-time ever more severely. The width of the escape cone (Figure 12) progressively shrinks. As it does so fewer and fewer light rays are able to escape. Ultimately only those at, or very close to the vertical can get away. Finally a critical stage is reached. At this point the cone closes, or more correctly, ceases to exist. From now on it matters not at all in which direction a ray of light travels. Come what may it can never escape! Neither can anything else, for the star is now "locked" within a gravitational prison. So far as the star is concerned the rest of the universe has ceased to exist—and, of course, vice versa.

Crossing this peculiar threshold does not imply that the fantastic escalation of the gravitational field will cease. Far from it. The gravitational field will eventually become of infinite intensity, crushing the matter of the star into zero volume. Infinite also will be pressure and density. It is at this point, with the unfortunate star crushed into "nothingness," that a "singularity" is formed.

The reader may be tempted to inquire what other odd physical manifestations might be expected in such unique and extreme circumstances. Well, of course, no one (to the best of our knowledge) has ever entered a Black Hole. And if they have, they have certainly never returned to tell us about it! Consequently we must content ourselves with predictions. Imagine a space vehicle approaching a Black Hole. It is now generally accepted that in a gravitational field time becomes retarded. As a result all chronometers in that space vehicle start to run slow with respect to our clocks and watches here on Earth. The closer that foolhardy crew approach the Black Hole the more slowly do their chronometers run. When eventually the craft reaches the event horizon, as far as we are concerned, they simply stop. Because of this, the approach of the space vehicle to the Black Hole, again so far as *we* are concerned, seems to take longer and longer. In fact the craft takes an infinite time to cross the event horizon by *our* reckoning. Members of the crew, however, find nothing amiss. To them watches and chronometers are behaving quite normally, for their physical processes, heart, respiration, brain, and general metabolism, have slowed down in phase. What then ensues we will come back to later since it has a distinct bearing on our main theme.

We are now in a position to predict how an embryo Black Hole would appear to external observers. The star concerned is collapsing under tremendous gravitational forces. Because of this the "cone of escape" is becoming steadily narrower and light rays increasingly less able to escape. To the eyes of external observers, therefore, the star is growing dimmer. We have just mentioned the fact that time (and hence clocks) becomes more and more retarded in an increasing gravitational field. In this respect the atoms responsible for light emission are a bit like clocks. They too "run" more slowly, causing a shift in the frequency of the emitted light toward the red end of the spectrum. So not only does our star become dimmer, it also becomes redder. Eventually it disappears entirely (Figure 14).

It would seem at a cursory glance that a spaceship crossing the event horizon of a Black Hole could well collide with the singularity, which is all that remains of the original collapsing star. On the other hand, how could it collide with something that has zero volume? This would represent a good question under normal

Figure 14
*Appearance of a star which has become a Black Hole.*

circumstances. Circumstances within a Black Hole are, however, very far from normal. It is thought that the question of collision arises only if a Black Hole conforms to two essential conditions: that it is *perfectly spherical* and that it *does not rotate.* Now it is most unlikely that a Black Hole would be perfectly spherical. For this there are a number of very sound physical reasons. Since an explanation would necessitate an excursion into the realms of mathematics we will ask that the reader accept this on trust. And what of rotation? We can only state that a *non-rotating* object in the universe would be a distinct oddity. Planets rotate, moons rotate, stars rotate—indeed, whole galaxies rotate (though they spend an inordinately long time doing it). Why, therefore, should a Black Hole be different in this respect, especially when the star whose collapse produced it certainly was not? In the circumstances we can safely accept rotation as fundamental. The two conditions are therefore satisfied. A spaceship can enter a Black Hole and it need *not* collide with the singularity within. Which is just as well since such a collision would be unpleasantly catastrophic!

Here for the record is the probable sequence of events likely to overtake any astronaut heading for a Black Hole. Very soon now he will cross the event horizon. As he does so his body will become subject to gravitational forces some thirty times as great as those he would experience at the surface of a *neutron* star, which, as we have seen, are not exactly negligible! As he approaches the center of the Black Hole the magnitude of these already massive tidal forces increases, resulting initially merely in the dismemberment of his *body*! The saga of his misfortunes is not yet complete. In quick succession comes disintegration of the molecules of the dismembered parts, the atoms constituting these molecules, the nuclei of these atoms, and probably in the end the basic particles of the nuclei themselves. By any standards, a spectacular if rather unpleasant demise, but over in a few milliseconds. Our foolhardy astronaut would, therefore, have no time to record his experience for posterity!

All this, we must stress, relates to an idealized Black Hole (perfectly spherical and non-rotating) and this, as we have already stated, is highly improbable. Let us, therefore, return to our more orthodox Black Hole. Here our astronaut will not, we believe, find himself on a collision course with a singularity. The picture has now completely changed and, in so doing, gives rise to fascinating implications with respect to space travel to other stars and other galaxies!

In Chapter 3 we saw that even a normal, well-behaved star has, because of its gravitational field, a region surrounding it in which space-time is mildly warped. In the case of a collapsing star/Black Hole the effect is infinitely greater (Figure 15). If, therefore, we envisage a region of space containing a number of collapsing star/Black Holes we might conveniently liken this region to a plain pitted by caldera-like depressions. Anything passing across this "plain" is quite likely to encounter one of these depressions and fall into it.

As the stellar collapsing process becomes ever more extreme, space-time curvature increases in phase. Now we come to the really fascinating possibility, for we have reason to believe that this

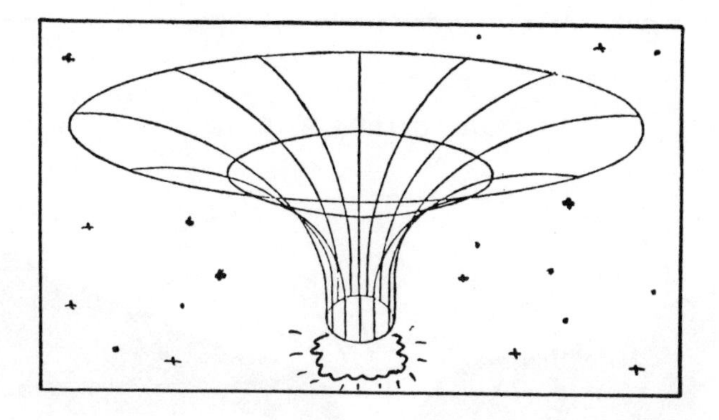

Figure 15
*Space near "collapsing" star curved by gravitational field of that star.*

curved space opens out into *another universe.* Imagine a mirror-image of warped space-time as shown in Figure 15. If the two are placed together (Figure 16) we have what is known as an Einstein-Rosen "bridge" (after the original postulators of this unique possibility). More often than not, it is referred to as a "wormhole," an appropriate if slightly inelegant title. The upper portion shown in Figure 16 links *our* universe with the event horizon; the lower, or mirror-image portion carries on into the other universe. The feature is also capable of a different (and perhaps more realistic) interpretation. The wormhole might conceivably connect two well-removed points in our *own* universe (Figures 17 and 18).

It is not our intention to go into the subject of other universes. In the circumstances a mention will probably suffice. In the preceding chapter we considered the possible shape of our own universe and stated that it could be of "saddle" form, comprising a skin of normal four-dimensional space-time enclosing five-dimensional non-space, all this in turn surrounded by yet more non-space. Beyond this it is possible, no doubt, that other, entirely separate universes might exist. However, for the present, a stage consisting of a single universe seems more than vast enough. We may be prepared to accept the idea of intelligent beings arriving from other stars, even from other galaxies. Perhaps it would be advisable for the present to ignore possibilities beyond these. Nevertheless, Black Hole theory does suggest possible links with other universes.

Talk of other universes and entry into them via Black Hole

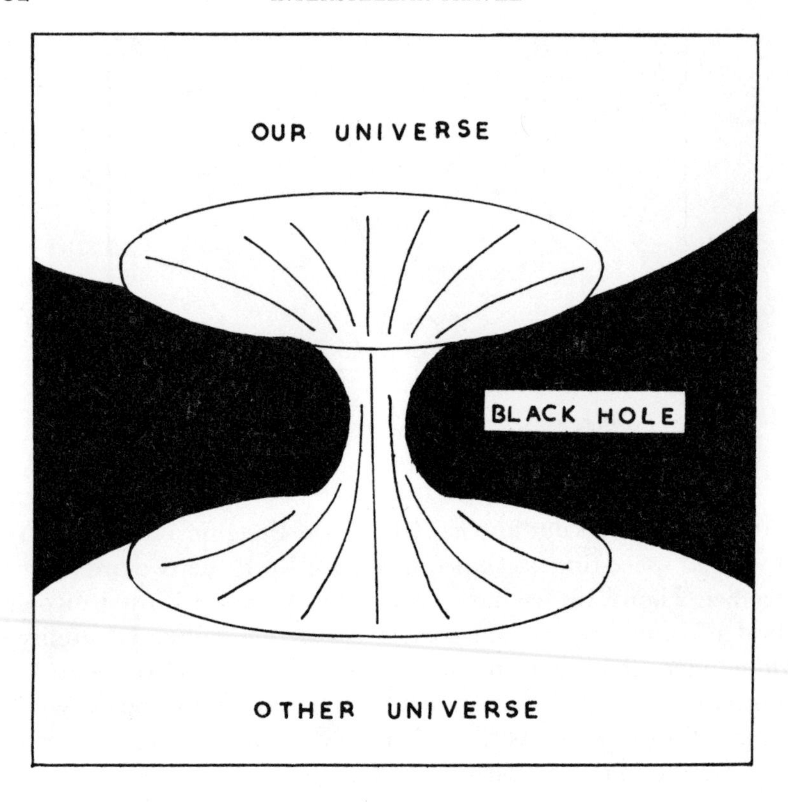


Figure 16

seems almost like an attempt to outdo science fiction. But if we accept the basic concept of the extreme warping of space-time by Black Holes, such things become theoretically possible. In this respect we have still not finished. There exists also the possibility of a journey from our own planet through a Black Hole which could lead us not to a far star or galaxy or even to another universe but instead back to an Earth either millions of years in the past or millions of years in the future! Farfetched, fantastic, impossible? No reader could really be blamed for thinking so, yet, in theory, a case can be made for these very things.[3]

Possibilities of this nature raise a number of paradoxical and intriguing questions. Suppose a strange, futuristic spacecraft were to descend on the surface of our planet tomorrow. We would assume, not unnaturally, that it must contain aliens, beings whose

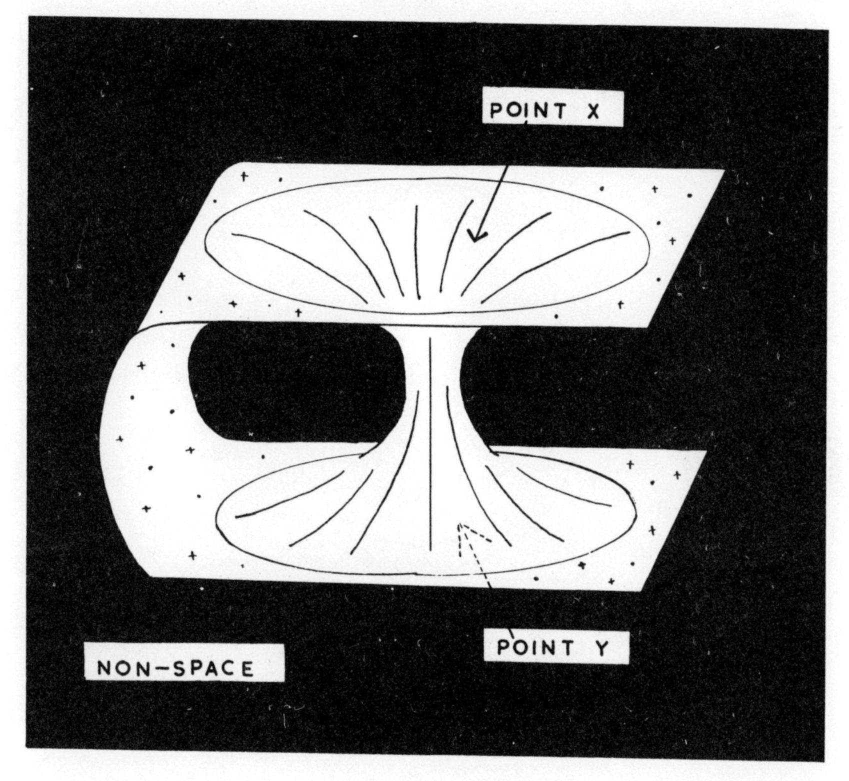


Figure 17

form could be very different from our own. But suppose, after what we have just said, that its occupants were instead terrestrials, fellow men and women of our planet's far future who, by virtue of space-time travel via a Black Hole, had decided to have a look at Earth in what to them would be its remote past! Time-travel, despite the earnest endeavors of science-fiction writers over the years, has remained a very difficult concept to comprehend. Though there exists an intertwined, almost indefinable relationship between space and time, it has always been easier to visualize space-travel than its temporal counterpart. Indeed, the mind will adapt with surprising ease to a monstrous extrapolation of the former while boggling unashamedly at the mere thought of being transported back or forward by even a few hours. The reason is not hard to find. Space is a dimension which is easily comprehensible. Time

Figure 18

may be real enough (for, after all, we are acutely aware of its passing—especially as we grow older!) but it is not an entity that can readily be visualized or defined.

Common sense (or what passes for it) also comes into the picture. An event with which we have been intimately concerned has taken place. It is almost impossible to conceive of a temporal shift which would result in our experiencing that self-same event again. Such an occurrence would be like the second run of a movie. We would know precisely what was coming and, had the event been of a less than pleasant nature, we would presumably seek to ensure that the circumstances which brought it about would not be repeated. This, of course, is just one facet of the problem, but it is probably sufficient to illustrate the magnitude of the difficulty. Time is an entity but one which, unlike space or a material object, cannot be seen or felt in the accepted sense.

It is important however to realize that space and time are inextricably linked. Defining this link is less than easy. Perhaps the best course is simply to state what, in a manner of speaking, is really quite obvious—that space (and material objects within it) cannot exist without time. The two are one—the space-time continuum. Time is in fact the fourth dimension, the others being length, breadth, and depth. Consider, for example, a block of some solid material. The normal dimensions, length, breadth, and depth, are clearly in evidence. The fourth dimension, time, is not, yet without time that block could not exist.

If, therefore, the distorting effect of near infinite gravitation, i.e., "Black Holes," can effectively warp space, then, since space is really space-time, time too should be distorted. In these circumstances the possibility of a "time-tunnel" may be as real as that of a "space-tunnel." The senses may rebel at the concept. This is natural enough. Nevertheless we have to accept it. Equally, the senses reel at the infinitesimal proportions of the atom, yet this we now accept with equanimity. In such matters it is not necessary that we comprehend, merely that we accept!

The time-travel aspect of entering a rotating Black Hole is, to say the least, intriguing. It is also full of complexity and paradox. In essence, the situation is as follows. A spaceship enters a Black Hole and, since the latter is of the rotating variety, the lethal singularity within is (hopefully!) avoided. The ship then emerges from the far end of the "space-tunnel" into another universe. Having achieved this, the occupants of the vessel might understandably have second thoughts and decide to retrace their steps. The remedy seems perfectly straightforward—simply allow the ship to fall back into the "tunnel-mouth" from which it has just emerged. At once they might find themselves confronted with a problem of mind-bending dimensions, for they could well emerge not into their own familiar universe but into yet another "other universe." The simple "worm-hole" could in fact be a *system* of "worm-holes" linking not just our universe with another, but with *several*! Indeed, some cosmologists claim that astronauts in this predicament could continue journeying from one universe to another without ever being able to regain their own. Though such an impasse could (theoretically) be avoided by selecting a Black

Hole of the nonrotating type, the disadvantage of having ship and occupants squeezed out of existence is not likely to encourage the experiment!

Now to the time-travel aspect. The occupants of a ship which has entered a rotating Black Hole and reached the point representing the center of the collapsed star have the option, if they are so minded, of going on a time journey. The only requisite is that their ship move in a circular path around the line about which the Black Hole is rotating and in a contra-rotatory direction. Each circuit around the rotation axis would gain for the astronauts an amount of time proportional to the spin of the Black Hole. If, then, they were able to regain their own universe, they could presumably put this to some advantage. If, on the other hand, they could not, then the exercise would be rather pointless.

Let us assume, however, that the former circumstances prevail. The ship would emerge not on this occasion into another universe but into the same one at a *different point in time*! The possibilities in these circumstances must surely be as bizarre and limitless as they are frightening.

Black Holes open up possibilities of immense scope. Many might be tempted to call them frightening—and who could blame them? During the past couple of decades astronomical research has brought to light much for which it is difficult to account in terms of presently accepted physical knowledge. We may at times believe we have explored the heavens pretty thoroughly, that we know most of the answers. Far from it. We have not reached the end. It is doubtful if we have even reached the end of the beginning.

In these last few pages we have thought predominantly of Black Holes as tunnels through space. The analogy is not unreasonable. But just as tunnels have their entrances so also do they have their exits. What can be said concerning the exit of a Black Hole space tunnel? For obvious reasons we cannot go and inspect the exit of a Black Hole in another universe. What, then, of Black Hole exits in our own universe, exits whose points of entry are Black Holes in far-flung corners of our own universe, or even in another? We must reflect that alien visitors could emerge from just such points.

To this we can only say that certain galaxies are known from which matter is *apparently* being ejected. Some cosmologists are already pondering over whether or not at the center of such galaxies there may lie the exit from a Black Hole and that at such points matter and energy are gushing up from other universes. These have been termed, perhaps rather appropriately, White Holes.

In this particular and highly conjectural field these are very early days. Nevertheless, cosmologists are becoming increasingly aware of the possibilities and implications. It would almost seem that we stand on the threshold of tremendous new vistas and it is tantalizing to reflect on what we may know of the universe a century from now, or more correctly, on what our great-grandchildren will.

So far we have done a great deal of theorizing. It must now be asked whether or not Black Holes are *known* to exist. To this question it is still rather difficult to give an unequivocal answer. As already stated, the X-ray source, Cygnus X-1 would *appear* to be a Black Hole/B class supergiant binary star system. The same can be said of the star Epsilon Aurigae. It is to double and multiple star systems that we should probably turn in our search because of the possibility in a number of suspected cases that a Black Hole constitutes one of the components, its presence being indicated by the gravitational effects exerted on the visible member of the system.

Thus far the Cygnus X-1 system would seem to contribute the clearest case for the existence of a Black Hole. NASA scientists using X-ray telescopes mounted in America's OAO satellite Copernicus have tied the binary supergiant system HDE-226868 to the Cygnus X-1 X-ray source and have detected evidence of the structure of the visible star's gas clouds swirling around and into the X-ray source (or Black Hole). Cygnus X-1 emits intense X-rays, the energy of which is a million times greater than the total energy output of our own sun. This possible Black Hole is estimated to have three times the mass of the sun but less than one-fiftieth of its size. The source of the X-rays is considered to be the mass of extremely hot gas falling into the Black Hole after being pulled off the visible supergiant star. The star could eventually disappear entirely down the Black Hole leaving no trace whatsoever of its former existence.

The mass of this Black Hole (at least three times that of the sun) has been determined as a result of the way in which the visible star moves in its orbit. Cygnus X-1 lies about 6,000 light-years distant and is believed to orbit its visible component every 5.6 days.

Let us now, very briefly, summarize what we have already said with respect to Black Holes and their possible potential as cosmic tunnels.

Current theory postulates that spherically symmetrical stars having masses greater than 1.4 times that of the sun must ultimately collapse. The gravity-resisting forces of the dying star are finally overcome by gravity so that the star collapses to a singularity, this being a point of infinite space-time curvature and infinite density. The singularity is surrounded by a region, known as a Black Hole, within which the forces of gravity are of such magnitude that all matter (including photons and therefore light) is unable to escape from it.

Matter falling into a Black Hole must be totally destroyed, rent asunder by tidal forces of unparalleled and unbelievable magnitude. This would be the fate of any spacecraft and its unfortunate occupants. In the case of a nonsymmetrical, rotating collapsing star the position would be different. Although the asymmetry cannot prevent the formation of a singularity, the in-falling matter (e.g., a space vessel) may avoid that singularity. In other words, the deformed Black Hole may act as a sort of tunnel or "wormhole" whereby matter entering it may be transported to another point in space-time, perhaps even into another universe.

The problems of interstellar and intergalactic travel with respect to Black Holes are as follows:

1. Black Holes would first have to be located and then identified as nonsymmetrical, rotating objects.
2. These might (almost certainly would) lie at vast distances from the "home" planet. They would first therefore have to be reached.
3. There might be no control over where such a "space tunnel" would lead a vessel.

All these difficulties might be obviated were an advanced alien intelligence able in some way to "create" a "synthetic" Black Hole at will in front of their ship. Equally important would be the ability to so orient it that the space tunnel thus created would lead precisely to that region in the universe selected as destination.

To think in such terms is to envisage a civilization about as far ahead of our own as ours is of the Stone Age on Earth. We have seen that the creation of Black Holes involves the manifestation of forces so great that the orthodox laws of physics are rendered invalid. It may indeed be that the artificial generation of such effects is completely and forever beyond the powers of any race irrespective of its cosmic maturity and technological skills. We simply cannot tell. For all we know, somewhere within this great universe of ours there may exist beings capable of something along these lines. Is there, after all, a strange irony about the space warps which several generations of science fiction writers have found so convenient?

It must be admitted that interstellar (or intergalactic) travel by means of a natural Black Hole, though theoretically feasible, could be rendered invalid by practical considerations. All likely Black Holes are incredibly remote from us (which, perhaps, ought to be a matter for gratitude) so that in the first instance one has to be reached. Moreover, though we have postulated conditions under which starships might pass safely through, this is all still theory and probably highly conjectural theory at that. Perils of an appalling and unknown nature might still confront such intrepid astronauts. In these circumstances, then, a technique leading to "space worm-holes" producible at will at any selected point in space would clearly be a better bet. Would a starship require a space-tunnel as vast as that provided by a Black Hole? Would not something of the same type, but of infinitely less magnitude, suffice? This, as we have said already, is something well outside terrestrial powers—but perhaps not those of a much more technologically advanced alien people.

The chapter on curved space speaks of a starship's being able to pucker and pierce the "skin" of "non-space," thereby penetrating that unknown, frightening dimension. This might be likened to piercing the skin of a *hollow* ball. The analogy in the case of a vessel seeking to adopt a Black Hole technique might be likened instead to piercing the skin of a *solid* ball and boring a tunnel through it to the other side. In this case, our vessel would not be entering the dimension of "non-space," but rather cleaving a tunnel of "normal space" through some other unknown entity. Admittedly, such concepts are highly speculative, yet there seems no valid reason why we should not try to envisage, however hazily, the possible parameters of a "universe" hidden to our senses.

While on the theme of small, artificially generated Black Holes and the ability of certain alien societies to produce these at will, brief mention should perhaps be made of their potential as possible weapons. Science-fiction writers and prophets of the future frequently talk of a "doomsday weapon." Over the years, this has taken many forms, but perhaps a Black Hole of relatively minute proportions, generated and directed at our world by alien beings, represents the ultimate. Sinking rapidly to the center of our planet, it would simply devour the Earth with cataclysmic fury. And lest we regard this as altogether too fantastic, it should be added that research teams in the United States, the Soviet Union, and Europe are already endeavoring to condense matter a millionfold to the density of white dwarf stars. Who knows where this process will stop—and who knows how far along this road some aliens have already gone! It is a sobering thought. There was a day not so long ago when Doomsday belonged only to the world of science fiction. Since $E = mc^2$ was given practical expression over the New Mexico desert, that day has gone. And, like departed youth, it will not return! In all this, however, we are digressing from our main theme.

No matter how technologically advanced and sophisticated a race may be, the irrefutable facts of time and distance render interstellar travel at best a one-way trip (and a lengthy one at that) so long as that race thinks purely in terms of the *conventional* universe. If such travel is to become a viable and really practical proposition a shortcut *must* be found, a shortcut in the most literal sense. We see today, hazily and dimly, that the physical universe may not after all be as it seems. We are beginning to realize that we comprehend it no better than the trout comprehends the great ocean lying beyond the mouth of its little stream. If alien vessels have in the past made cosmic landfalls on our unimportant little planet they have come not from its sister worlds about the sun but from cousin worlds, from worlds of other suns. Their occupants and the great civilizations which spawned them must therefore in some way have been able to bend the seemingly immutable laws of time and distance. They must have found a shortcut, a cosmic detour. Is that detour a "tunnel" in the space-time continuum? And if in the past they have used such a technique to reach us, may they not do so again? Is that far star whose light has taken several hundreds of years to reach us so remote after all, or is there a strange, short tunnel at the end of which it lies beckoning?

It may be that shortcuts between stars due to space curvature, with which we dealt in the preceding chapter, have even greater relevance with respect to Black Holes. For long it has been scientifically fashionable and correct to dismiss interstar travel. "The stars are too far," say the pundits. But are they?

# 5. TIME DESTROYED

It will be appreciated that "conventional" systems of interstellar transit along the lines of generation travel or suspended animation can be only one-way trips. This must apply to alien visitors just as much as it would in future ages to our own kind. In the absence of more sophisticated and practical techniques these things would have to be accepted, at least until a more advanced alternative became feasible. Creeping colonization and alien dominion may even now be under way using these relatively slow and imperfect techniques, a form of encroachment which could, for all we know, be getting very close indeed to the environs of our own sun and solar system.

There exists a distinct possibility, however, that an entirely different alternative could have been selected and perfected. This would to some extent continue to render long-distance interstellar voyages something of a one-way affair, though for a vastly different reason. It would, however, remove from them the long weary pathos of generation travel. The concept is one which has intrigued, perplexed, and tantalized minds for years and is now regarded as still something of a paradox. It is the idea of "time dilatation" travel—in a peculiar sense a form of time travel—but *only* into the future.

It has been suggested and indeed by now has been fairly well proven that if an object be given a velocity approaching that of light (186,000 miles per second or 300,000 kilometers per second) distance can, in a sense, be greatly telescoped. First, perhaps, we should look at this in a popular and essentially lighthearted way. Suppose a space vessel were to travel to our nearest galactic neighbor, the Great Galaxy in Andromeda. This great star system, similar to but larger than our own Milky Way, lies some 2¼ million light-years from us. Suppose also that it were possible to impart to this vehicle a velocity very close to that of light, say 0.99c (c being light velocity). Our galactic starship would reach that other galaxy in Andromeda, so far as the computations of its occupants were concerned, in something like 27 years. And if having finally got there they decided the place wasn't much to their liking and desired to return, the homeward leg of the trip would also (apparently) take 27 years. All aboard would be about 54 years older than when they left Earth but they would at least have had a fascinating and enthralling trip. But as finally they spiraled in through the outer stars of the Milky Way toward the sun and solar system they would make a strange and, we imagine, highly disconcerting discovery. The Earth they were now approaching would no longer be the Earth they had all (apparently) left 54 years ago. It would look considerably different, one of the more obvious points of difference being the fact that the continents had quite clearly shifted. As they moved steadily closer they would notice many other respects in which Earth now seemed different. It would not just *seem* different, it *would be* different for, while they had voyaged to Andromeda and back, Earth would have aged by something like 4 million years. Fantastic? Well, at first glance so it might easily appear, but a mathematical case can be made for it and to some extent even a physical one.

In this example we have gone well out on a limb by selecting not just another star but another galaxy. If instead we consider travel between stars in our own galaxy the strange disparity in transit time is considerably diminished. Let us think of a typical example. A journey to the nearest star, Alpha Centauri, and back would, to the occupants of the space vessel concerned, take 9.4 years if the velocity were two-thirds that of light velocity (200,000 kilometers per second). To those remaining behind on Earth the time would be 12.6 years. In such bizarre circumstances starship crews would appear to their terrestrial-bound friends to have discovered the fabulous and legendary Elixir of Youth.

It must be stressed at this point that the effect, assuming it is not just a mere mathematical paradox, is largely dependent on the attainment of velocities as close as possible to that of light. A velocity of, say, one-fiftieth that of light, though tremendous by present-day standards, would not really make a very great deal of difference. It is in fact essential, if time is to be telescoped or "destroyed" in this fashion, that velocities in the range 80-90 percent that of light are achieved. So far as we on Earth are concerned, for the present and foreseeable future the matter must rest since there is as yet no power source or technology in sight which would enable us to attain velocities of anything approaching this order.

Once again, though, we have to stop and consider what may have been achieved by other older and more mature civilizations within the vast galaxy of the Milky Way, by peoples as much ahead of us as we are of terrestrial Stone Age man. If there are those who have achieved such a capability, then it is not impossible that Earth and the other planets of the solar system are now within reach of their far-ranging starships. Presumably, the lack of aging would affect them too. On return to their home planets they would find relatives and friends always that bit older. And if their space odyssey had been a long one their one-time contemporaries would in all probability be dead. To an unemotional race this might matter little, if at all. If the aim was simply one of colonization (a one-way trip), presumably such an issue would not arise. Colonization of other planetary systems by alien beings must, of course, for us have sinister undertones, for reasons which should be only too obvious!

Before proceeding further we must take a closer look at this concept of time dilatation and ascertain, if we can, what relevance it has—a relevance that might apply to us a few centuries hence, or *tomorrow,* if aliens used it as a means to intrude upon the solar system.

Like space curvature and Black Holes, time dilatation, especially at a first glance, seems a bizarre and highly unreal concept. This is probably just another manifestation of the fact that the physical universe is not as it seems.

A proper explanation of time dilatation involves the use of a certain amount of mathematics. The pages of a book of this nature, however, seem hardly the place for algebraic formulae and equations no matter how straightforward. To some extent, then, the basics of the idea must be taken on trust. Readers wishing to see a mathematical treatment will easily find a number of suitable books on the subject.

Essentially the position is this. A car going from A to B will, to its occupants, take *less* time to cover the distance than it does to stationary observers. The difference, however, is infinitesimal and could not normally be measured. The same can be said of a jet plane, even of a supersonic transport like the Concorde. The speeds involved in this case, though considerably greater, are still not sufficiently high for the difference to be tangible. When, in 1963, Major Gordon Cooper of the United States Air Force made his 22-orbit spaceflight around the Earth it was estimated that the difference was a *millionth of a second*—and this for a velocity of around 20,000 miles per hour. The discrepancy becomes apparent only when the speed of the spacecraft comes very close to that of light itself, generally denoted by the letter "c" (186,000 miles per second or 300,000 kilometers per second). Obviously we have a very considerable way to go before we achieve velocities of this order.

Let us consider a space voyage from here to the "close" star Alpha Centauri (4.3 light-years distant) at a velocity of c/50, that is, 3,720 miles per second. By simple arithmetic the time for the journey should be 225 years. In fact, due to time dilatation this is going to amount to two months less. Two months in 225 years is not much of an advantage. Indeed, it would barely be noticed. Velocity is still too low. Nevertheless, the prospect, if the velocity of space vehicles can be greatly increased, becomes very interesting indeed.

The inevitable question arises: Is all this real or is it merely an intriguing mathematical paradox? Is there in fact any physical substantiation for something which seems almost an affront to logic and reason? Apparently there is.

Time dilatation is obviously not going to be observed in the case of running men, running dogs, cars, trains, jet planes, even Apollo moon rockets. It is not going to be noticed even with our fastest-moving space vehicles to date, though of course the degree of such dilatation *is* rising as we consider objects with greater and greater velocities. It can, however, be observed and measured in the case of certain fast-moving subatomic particles known as muons. Muons reach the Earth from outer space and possess a wide range of velocities. Experimental work in this field certainly appears to confirm the validity of time dilatation.

The argument over asymmetrical aging (that is, normal aging of the Earth-bound with respect to the slower aging of the starship travelers) has now continued (often very heatedly) for several decades. As might be expected, many have found the idea of unequal aging very difficult to accept. More recent experimental work in this field still appears to confirm the overall validity of the concept. A recent experiment carried out by H. Keating and R. Hafele, in which an extremely accurate "caesium" clock flown around the Earth was compared to a stationary clock, confirmed that the moving "caesium" clock *did* record the passing of *less* time than the stationary one.[1,2]

The occupants of a starship moving at 0.99c (very close indeed to the velocity of light) would traverse a distance of one light-year in approximately 0.14 year of "Earth time." This in fact gives to them an *apparent* effective velocity of roughly 7c with respect to Earth—seven times the speed of light, or 1,302,000 miles per second! So in a sense the hyper-light drives of science fiction are at least a theoretical possibility. The Earth-bound observer, of course, sees this distance covered in just a little over a year.

This all seems very tidy and perfect, but readers smitten with a sudden (or not so sudden) desire to leave Earth for the stars should reflect on a number of problems, two of which predominate:

1. The question of asymmetric aging. If the star of one's choice is fairly remote, then it would be better to settle for a one-way ticket. A return would hardly be meaningful!
2. The energetics of the situation must always be borne in mind, for to take advantage of this principle involves accelerating a starship from rest to relativistic velocities (that is, velocities near to the speed of light).

To some, the first disadvantage might be acceptable—it is merely a question of choice—and probably circumstance. So far as the second is concerned, choice has nothing to do with it. Here hard reality rules and in this instance it is particularly hard. It need hardly be stressed that the attainment of near-light velocity is nowhere in sight so far as we on Earth are concerned at the present time nor does it seem even remotely likely in the foreseeable future.

An increasingly common subterfuge among some science writers of late has been to produce a graph of speed increase against time starting with the speed of the horse in the nineteenth century. Such a graph extended to the present time is virtually an exponential curve since starting at horse velocity it continues to the speed of modern space probes. So far this is fair enough. The increase has been spectacular and dramatic enough by any standard. Unfortunately there is a temptation, admittedly a very strong one, to extrapolate such a curve to the year 2,000 and beyond. When this is done we find that light velocity is likely to be attained around the turn of the century, which is clearly unmitigated rubbish. There is about as much sense in this as expecting a sick person with a rising temperature to ignite! The gradient of such a time-velocity curve must eventually level off.

Some other galactic societies much older than our own will probably have attained near-light velocities. Assuming therefore (and about this we always have to be cautious) that the concept is not merely a mathematical paradox (which certainly it doesn't seem to be), then starships could in effect destroy both time and distance and thereby reach other solar systems, including our own. In a sense the approach seems almost orthodox compared to "curved space" and Black Hole travel. Moreover, asymmetrical aging is not going to be highly significant if our alien visitors have to traverse only a few light-years.

An increasingly popular science fiction theme is that of "teleportation," in which the molecular fabric of spaceships and their occupants is reduced by electronic means, transmitted, then received or "reassembled" at the other end. It seems very unlikely that such a sophisticated and effortless method of space travel will be available to us terrestrials for a very considerable time. There are reports at the present time of research in this direction both in the United States and in the Soviet Union. The basic thinking behind the idea is that of the "bio-pack"—that is, the breaking down of spaceship and crew harmlessly (it is hoped!) into pure energy components which constitute part of an electrodynamic bio-pack which can then be sent into space on a radio beam directed toward a star having a system of planets one of which is considered able to support the type of life-form involved. On arrival several years later (for radio waves move with the speed of light) spacecraft and crew could hopefully be reconstituted as going concerns!

Clearly such a scheme, if feasible at all, could be carried out only by beings very much more technologically advanced than ourselves. The idea represents yet another possibility for bridging the immense interstellar gulfs. The reason for including it at this point is that such a means of transit seems to suggest a fresh paradox not entirely unrelated to that of time dilatation.

The paradox, if it is a paradox, is this. We have already seen that travel at or approaching the velocity of light invokes time dilatation. Teleportation, in essence, is travel at the speed of light. In the circumstances, then, should the effects of time dilatation show? We will leave the reader to think about it. No doubt, along with most, an experiment in teleportation is not one in which the writer would like to be personally involved!

Time travel has, ever since the late H. G. Wells wrote his memorable *Time Machine,* been a favorite theme in science fiction. Always, however, it has been seen as a patent absurdity, quite incapable of realization. So long as we think in terms of a specific device capable of transmitting persons, animals, or objects at will into the future or the past this verdict is likely to stand. Time dilatation resulting from travel at relativistic velocities is quite another matter. Here there may be potential for journeys into the future. Such journeys will be journeys with a quite different purpose—the purpose of star-hopping. To purposeful, inquiring, and ambitious alien beings this may rank as the most important and vital reason of all!

# 6. TACHYON TRAVEL

Until now we have continued almost automatically to invoke the generally accepted belief that it is impossible to attain the speed of light, let alone exceed it. Since we are (and for long must remain) unable even to approach it, it is clear we cannot yet put the hypothesis to the test. According to the appropriate equations, a velocity equal to or in excess of that of light makes no sense, mathematically at least. It is reasonable to assume therefore that this confirms the belief.

Nevertheless, it has been suggested from time to time that this need not necessarily be so, that we are as yet unaware of the true position, that the mathematics is the product of a false premise and because of this points in the wrong direction.

From the standpoint of interstellar travel, hyper-optical velocities, were it not for the restraints of Einstein, might be a good thing—a relatively simple answer to a difficult problem. If we desire to proceed from A to B on Earth, transit time depends only on the speed with which we travel. An analogous state of affairs with respect to a journey from star A to star B, or even galaxy A to galaxy B seems a neat and tidy answer.

Hyper-optical velocities were an early favorite with writers of science fiction. It enabled them to surmount the time and distance barrier in a way that seemed a little more orthodox than recourse to time warps and the like—all in a very relative way, of course! In the circumstances the word "relative" is highly appropriate. The ship simply went into "hyper-drive" and that was that. If the writers were aware of the restraints imposed by Einstein and time dilatation they cheerfully ignored them. (On reflection who could blame them!) Nevertheless, there was at times a recognition that peculiar things might happen at such fantastic velocities. One example stands out plainly in the memory of the writer, dating back to 1939. Title and author have unfortunately been long forgotten. Even the precise details of the story are now more than a little hazy but, as near as can be recalled, Earth and the entire solar system face impending and overwhelming doom. As a result it has become vital to search among the stars for a safe haven suitable to the human race. On the ship first able to exceed the velocity of light very odd things do indeed begin to occur. Due to an inexplicable molecular rearrangement sugar suddenly loses the property of sweetness and because of a spectrum shift colors change in a weird way. The only girl aboard is reduced to near hysterics when she discovers that purple lips cannot be restored to their natural hue by lipstick that has turned equally purple! She gives up the uneven fight when she realizes her auburn hair has changed to emerald green.

Apart from time dilatation it is almost impossible to predict what peculiar physical effects might occur were velocities in excess of that of light ever to become attainable. Presumably, collision with even the smallest grain of matter would prove catastrophic—unless relativistic effects sweeping scythelike ahead of the fleeting ship cleared its path.

It must be admitted that so far the evidence remains quite heavily weighted in favor of light velocity representing a limiting factor. Such evidence as there is to support the opposite viewpoint is slender in the extreme. Nevertheless, we cannot, in all fairness, pass on without taking a closer look at the situation as it presently stands. For all we know, a few millennia ago certain alien races somewhere in the galaxy did just that and, because they did, are now able to roam the galaxy with considerable freedom.

In fact what relativity theory *really* tells us is that a velocity *equal* to that of light is impossible. It does not deny the possibility of speeds *greater* than that of light. This is clearly a different matter. In the circumstances it is probably unwise to place too much reliance on mathematics that *might* just be a siren song leading us onto the relativistic rocks.

Let us for a moment or two forget about the so-called light velocity barrier and assume for the sake of argument that beyond it velocities greater than that of light *are* possible. This implies that, in a manner of speaking, another "universe" lies there, one comprising particles unable to travel *below* the velocity of light (not to be confused with those other universes we spoke of in Chapter 4). If such particles exist they must always be traveling with velocities in excess of that of light. We simply cannot detect them and consequently we have no means of confirming their existence. Thus we do not know whether they are there or not. Oasis or mirage? This is really what it adds up to.

Our hypothetical particle has nevertheless been given a name: the tachyon.[1,2,3,4] We will return to it in a few moments. But first, a brief digression is relevant at this point.

Reference can be found in radar technology to objects known as wave guides. These are rectangular tubes fashioned from either copper or aluminum along which high-frequency radio waves pass. In so doing these waves create electromagnetic patterns which also pass along the wave guides but at speeds *in excess* of that of light. Unfortunately, these patterns cannot be used as the basis of a signaling technique. This would involve *changes* in the patterns and such changes, it can be shown, move only with *sub-light* velocities. Perhaps someday a basis may be found here for an interstellar communications network whereby electronic messages, unhindered by the long delays inherent in conventional radio, may be exchanged between star systems. Admittedly at the moment this seems improbable. What this fact does emphasize, however, is that here at least is something which *does* exceed the speed of light. It is not energy and it is not matter. It is in fact only an appearance, a configuration. And, incidentally, it does not invalidate Einstein, though the mathematics proving this are highly involved, as might be expected.

And so now to the mysterious tachyon. Is it fact or fiction? Tachyons are defined as particles moving with infinite velocity which, when approaching the velocity of light, are *slowing down.* This certainly represents a unique change in the accepted order of things. According to physicist Gerald Feinberg, tachyons might have velocities as high as a *billion* times the speed of light.[5] What, it may well be asked, happens to tachyons when they reach the speed of light? The answer, according to the theorists, is delightfully simple: They cease to exist!

It cannot be too strongly emphasized that at the moment tachyons are merely an abstraction. Nevertheless, attempts to confirm their existence are currently proceeding. It need hardly be added that this is no easy task. The basic difficulties speak rather loudly for themselves. Here are (so we suppose) particles that are infinitely small, traveling with velocities that can only be described as incredible. Since they are traveling beyond the speed of light we cannot see them, and if we slow them down sufficiently they cease to exist. The only feasible approach to the problem is to try and detect them by virtue of their effect on some entity we *can* observe. A major complication is the fact that it is difficult to attribute any effect or effects categorically to the presence of tachyons.

The approach to the problem so far has been to use a hollow sphere made of lead, the interior of which is virtually a vacuum. Provision is made whereby any element or elements contained residually therein can be bombarded by high-energy particles. Theoretically, tachyons, if they are produced, will flash across the sphere at supra-light velocities and in so doing produce a distinct bluish glow known as Cerenkov Radiation. Such an effect, according to theory, could arise in a vacuum only from bombardment by particles moving with velocities greater than that of light. So far, it must be admitted, Cerenkov Radiation has been apparent only by its absence, but this is no real reason for supposing that tachyons do not exist. The experimental technique may simply be inadequate, in which case another (very probably others) will require to be devised. This will be no mean feat.

Latest information to hand regarding tachyon research is a little more promising although it is still too early for wild optimism. At the University of Adelaide in South Australia two physicists, Roger Clay and Phillip Crouch, have recently reported something that registered on their cosmic ray detectors well in advance of subsequently recorded cosmic ray showers. This, they believe, may constitute the first evidence of tachyon existence. Prior to this it had been postulated that tachyons might be produced within the atmosphere by cosmic ray showers. Since by definition these would be traveling in excess of light velocity they must reach the surface of the Earth before the shower itself. If capable of producing a response in ordinary particle detectors it should, as a consequence, be possible to observe the presence of particles associated with the cosmic ray shower arriving some tens of microseconds earlier.

Clay and Crouch in reporting the event are careful to retain a tight hold upon the reins. Following a statistical analysis of their measurements they merely stated that they had "observed non-random events preceding the arrival of an extensive (atmosphere) cosmic ray shower."

Now we really come to the sixty-four-dollar question. In what practical way might alien beings make use of tachyons? Here again we can only grope. Anything like a precise or definitive answer at this point in human affairs is clearly out. Were it otherwise we ourselves might be starting to think in terms of tachyon-driven space vehicles with the entire galaxy as our stamping ground. For the present and foreseeable future any such vessels appearing in our skies must have come from worlds whose sun is not our sun, whose occupants may or may not look like men!

If an advanced galactic community has succeeded in harnessing the power and potential of the tachyon, an appropriately designed and constructed starship could then be accelerated to the speed of light, at which point the tachyon drive would take over. There is a rather obvious snag here which we might conveniently call the "velocity gap." A "conventional" drive takes the ship up to 0.99c—that is, almost, but not quite, the speed of light. The tachyon drive takes over just beyond c (the speed of light). There is thus a quite definite gap. Bridging this gap is, of course, the problem.

Now if one is to proceed from a velocity less than that of light to one in excess, it seems reasonable to conclude that in so doing one must pass *through* the speed of light—that for a very brief instant velocity will equal c. To deny this seems simply an insult to common sense. If we take an automobile from rest to 70 miles per hour we expect that at one point the speedometer needle will indicate 50 mph. It is certainly hard to envisage it going from 49 to 51 without passing 50 on the way!

Despite the logic of this, it *is* just possible that in the circumstances we may have to accept something along these lines no matter how ridiculous it may seem. Somehow that critical velocity "c" must be bridged.

The answer is thought to be bound up in the strange realm of what is known as quantum mechanics. Without fully going into that here, we can nevertheless invoke the principles. A useful analogy is the way electrons orbiting the nucleus of an atom can jump from orbit to orbit if there is a change in their energy state.

In so doing they emit pulses of energy ("quanta"). Electrons so affected, it is postulated, leave one orbit and appear in the next *without* traversing the intervening distance. Dare we suppose that some alien races, thanks to the acquisition of ultrasophisticated technologies, have been able to bridge this gap—to render their propulsion systems capable of acting in the fashion of energized orbiting electrons? Familiarity and facility with tachyons may even have permitted the evolvement of techniques whereby the elusive particles have been "persuaded," however briefly, to exist at velocities marginally less than "c." For us at present this can be only purest speculation.

In science the last word is never spoken. It is doubtful if Einstein's theory of relativity represents the ultimate in physical thought. There are already those who cast doubts upon some of its facets. Einstein has presented us with a new and incisive instrument with which to probe the universe, but it is doubtful if this, or any other, will ever prove wholly adequate. Perhaps the velocity of light is not after all a limiting factor. Indeed, has it always been 186,000 miles per second—and must it always be?

There are doubtless other practical difficulties and dangers which beings traveling at hyper-optical velocities would have to face. If at points amid the far stardust of the galaxy some have tamed the wild beast that is hyper-optical propulsion we may feel quite sure that in blood (or whatever the alien equivalent of that may be) they must have paid a very high price indeed—as will we, if ever we reach this stage in technology.

As things stand at present, travel at near-light velocity, it seems, must constitute a form of time travel.[6] In the preceding chapter we quoted the example of a round trip by starship to the Great Galaxy in Andromeda which, for its occupants, would involve something on the order of 54 terrestrial years, though in fact Earth would be 4 *million* years older. This clearly represents a jump, and a monstrous one at that, into the future. It is impossible to envision what the equivalent effects would be were the same journey carried out at a hyper-light velocity. Whatever validity there may be for travel into the future, it is not possible to make any case for travel into the past.* There would seem to be absolutely no risk of return from an

* In the case of Black Holes this may not be so.

interstellar or intergalactic space voyage to an Earth still in the Stone Age. This can be seen as either fortunate or unfortunate according to one's viewpoint and frame of mind. At the present time such a return is not without attraction!

Beyond this there is little further that can be said at the moment. Tachyons *may* exist. Present indications are mildly promising. If they do, then one form of matter at least is able to exceed the speed of light. This would undoubtedly raise many interesting and intriguing questions. Indeed, the question mark hanging above "c," the so-called limiting velocity factor, would probably stand out just that little more boldly. Whether particles of so fleeting and elusive a nature would be harnessed is impossible to tell. Could matter in the shape of a starship be sent hurtling through interstellar space at these fantastic velocities without dire and probably very peculiar physical effects taking place? Certainly our science fiction heroes and heroines have been doing it for years and getting away with it, but the thing in practice, we suspect, would be a very different matter.

And yet, perhaps beings on remote planets in far-distant times once thought these same thoughts and penned similar words. They continued the search, tracked down the elusive particles, and gained a new insight into the nature of the physical universe. In time they built ships to sail the ocean of stars. There were tragedies—ships disintegrated or dissolved into fiery mushroom clouds. They held fast to the dream—and the dream became in the end reality. If that reality should ever approach our world, let us hope and pray it will not attain the dimensions of a nightmare to our people!

---

Let me relate a personal sidebar to the tachyon story. In early 1975, a silver, disc-shaped craft began to be observed around Hinwel, Switzerland, by a local farmer named Eduard Meier. Over the years he saw and photographed more of these crafts, and recorded the purported sound of their propulsion units. Both the photos and tapes have been rigorously examined by American experts, who could discover no evidence of faking.

Meier subsequently claims to have conversed with occupants of one of the ships, notably one who claimed that they originated from a planet in the Pleiades star cluster some 500 light years distant.

In 1982 I was visited by senior members of a scientific team investigating the story. I find no reason to doubt their expertise or their sincerity, but more than that I cannot say. The matter has been the subject of a film entitled "Contact," as well as two books.

# 7. NO STAR TO GUIDE THEM

So far we have concerned ourselves only with the *techniques* aliens might adopt to traverse the galaxy. However, a transit system can be effective only if there is a suitable *guidance* system as well. We must therefore think a little about this aspect but only in terms of conventional space. Guidance with respect to curved space and Black Holes will possess quite unique parameters of its own and it is virtually impossible for us to think in such terms either now or in the foreseeable future.

Interstellar navigation is clearly a necessity to any spacecraft that ventures out into the dark, empty cosmic ocean beyond the outermost planets of its own system. This will apply with equal emphasis to Earth starships of the far future. For the present, however, it must be regarded as a wholly alien preserve.

It might seem that there is virtually no problem in proceeding from star to star. At sea, land is soon lost to sight and the problem is apparent. Surely in deep space there can be no problem with the destination star clearly visible? An alien starship leaving a world of Alpha Centauri for the planetary regions of our sun is merely directed by its commander toward Sol and kept on that heading. Basically this might be true in the case of a "generation" or "hibernation" type of ship on a one-way colonizing mission. Time in these circumstances is less vital. Such ships will plod along at relatively low velocities, probably of the order of 2 percent of the speed of light (2 psol), taking perhaps centuries to reach their destinations. Civilizations that have not as yet found a means to utilize curved space or Black Holes will most likely have developed and refined the technique of relativistic (that is, "time dilatation") travel. Such ships will travel at high sub-multiples of the speed of light to gain as much advantage as they can from this strange and seeming paradox.

Now inconvenient and disconcerting though it may be, something very odd is going to happen in these circumstances.[1] It will commence at about the moment the aliens' starship reaches a velocity of 37 psol. At that point their own sun, Alpha Centauri, is going to disappear from sight and just for good measure their destination star, our sun, is going to do precisely the same thing. Clearly the navigators are going to have problems, problems undreamed of by marine navigators.

Navigation at sea is based on the star constellations and especially on a number of selected bright stars. When atmospheric conditions are poor these bright stars can often still be discerned. In deep space, where there is no haze, the stars, even the fainter ones, stand out much more distinctly. As a consequence the familiar constellations are less obvious and identification rendered that much more difficult.

The marine navigator is called upon to be familiar only with one set of constellations. Not so the starship navigator. As he traverses light-years of space the configurations of the constellations gradually change. A few hundred light-years out from the sun the pattern of the stars will be very different from that to which we have grown accustomed. Celestial navigators will therefore require either star charts to cover an infinity of different situations or a set of galactic coordinates enabling them to fix their position with respect to their home star (Figure 19).

At fairly low velocities (up to 10 psol) there are no great or insuperable problems. To obtain his position the starship navigator merely selects a trio of widely separated bright stars whose coordinates are known in relation to the "home" star. By using a special sextant and adjusting until all three stars are superimposed in its sights he can then determine the angles they subtend with each other. From specially prepared tables he can then read off the distances corresponding to these angles and thereby determine his position. A few months later he repeats the operation. He is then able to calculate the distance covered and from that the mean velocity of the starship. Slight corrections in course will occasionally be necessary. Such changes, however slight, involve

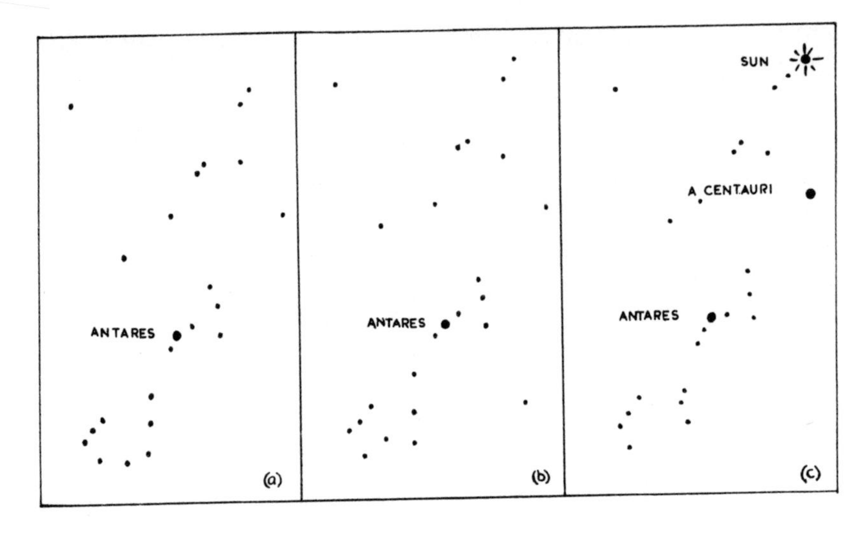


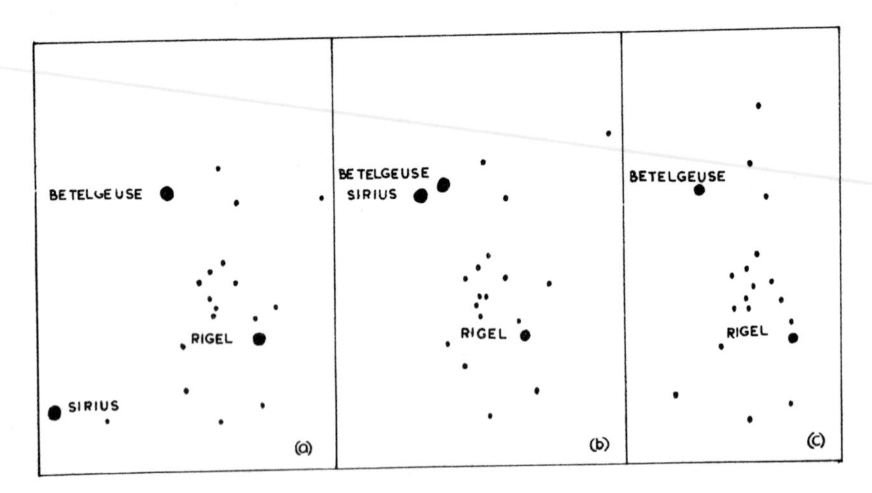


Figure 19

*(Top) Scorpius and adjoining constellations as seen from (a) the Sun (b) Alpha Centauri (c) Epsilon Eridani*

*(Bottom) Orion and adjoining constellations as seen from (a) the Sun (b) Alpha Centauri (c) Epsilon Eridani*

changes in velocity and the navigator must afterward recheck to confirm direction, velocity, and acceleration. There may also be occasions when major changes in course are necessary, for example, to avoid a cosmic dust cloud. A busy time lies ahead of the navigator though doubtless our alien friends have sophisticated computers to aid them. It will be necessary for him to determine the magnitude of the new speed vector and compound this with the former. Space is a frictionless medium. Because of this, acceleration, and time spent accelerating, result in velocities that are strictly in accord with mathematics.

The only real danger likely to accrue from indifferent navigation is that *overcorrection* might sometimes occur thereby depleting fuel stocks to a greater degree than planned. This could conceivably result in a perilous situation—insufficient fuel either for further course corrections or for braking purposes on reaching the destination star. With the degree of sophistication expected of advanced aliens this type of thing would presumably not occur.

The real difficulties arise when "relativistic" velocities are involved. Let us return to our imaginary alien vessel heading for the solar system from a world of the star Alpha Centauri at a speed very close to that of light. No longer do the "simple" conditions outlined above prevail. The alien vessel has "locked" its sights upon our sun and is now well under way, its velocity gradually mounting. The first hint of something unusual comes when the ship reaches 15 psol. Then the stars lying directly ahead gradually start to become brighter, whereas those lying directly astern show a spectrum shift toward the red—that is, yellow stars become orange, orange ones red, and the original red ones disappear. This is a clear indication that the ship and its occupants are moving ever faster toward the stars ahead and ever faster away from those astern.

Worse is to follow. When the ship reaches 25 psol it appears as if the stars ahead are clustering together. The transition is gradual but none the less real. Because of the apparent distortion of the constellations star charts are rendered useless. The position astern is reversed. There the stars are apparently thinning out and growing dim. Alpha Centauri, home of the aliens, is now barely distinguishable. Such phenomena are a sure sign that the ship is now well inside the strange realm of relativistic velocities.

At around 36 psol Alpha Centauri finally disappears entirely to be replaced by a small but growing circle of darkness. Even worse, especially from the navigator's point of view, our sun, the star toward which the ship is hurtling, has also disappeared and been replaced by an expanding circle of darkness. Toward the sun the spectrum shift is toward the violet end. The sun's light is now within the ultraviolet region and therefore invisible to the eye.

When the ship achieves 50 psol navigational problems are further compounded. Ahead of the alien vessel the dark circle has now grown to a huge cone. The position astern is similar but here the cone is of even greater extent. Although the ship is in interstellar space surrounded by stars the only ones now visible are crowded together in a sort of cosmic "barrel" around the ship. Nor is this all, for it is as if these stars had formed a spectrum of their own. Toward the front—in the direction in which the starship is traveling—the stars are a brilliant bluish-white. The colors of the stars change gradually through green, yellow, and orange until at the rear of the "barrel" they are red [2,3] (Figure 20).

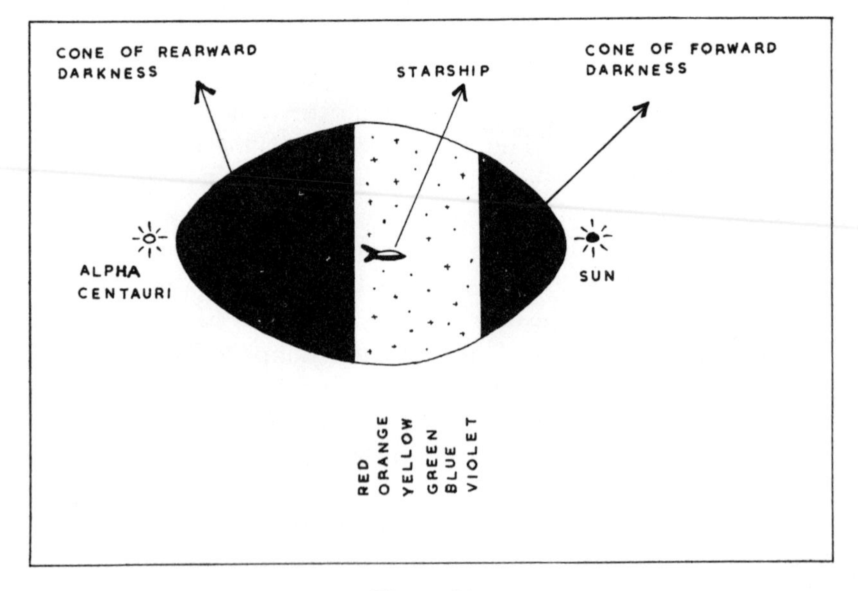


Figure 20

In circumstances such as these it would be virtually impossible to distinguish a constellation. Even if this could be done a further impending change in the surroundings would render it pointless. As velocity mounts still further the "ends" of the "barrel" begin to move upward then forward together. Soon no stars at all are visible in the hemisphere facing the stern of the ship. The forward cone of darkness is now also beginning to contract. As velocity further increases both ends of the barrel can be seen in the hemisphere ahead of the ship. By now the band of stars has been reduced to a mere ring of color-banded packed stars (Figure 21).

The difficulties of navigation no longer speak for themselves—they shout! Now the alien astronauts know neither where they are nor in which direction they are going. To determine these things they must slow down, a protracted, fuel-consuming business.

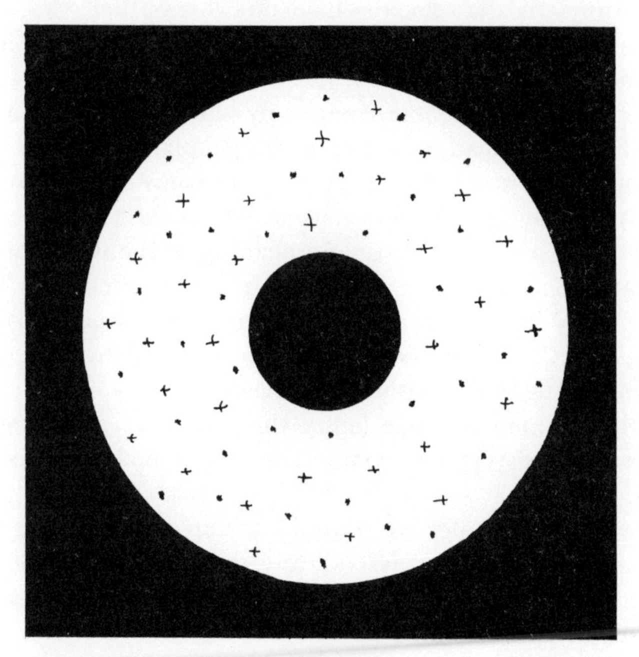

Figure 21
*As seen ahead of a starship, the effect of attaining 99 percent of light velocity is a stellar "rainbow."*

Having made the necessary observations and calculations and returned to course they must then reaccelerate (also protracted and fuel-consuming) in the sure and certain knowledge that precisely the same thing is going to happen, again and again, ad infinitum.

Should by any chance the alien astronauts see the "rainbow" ring of stars move to one side and assume a crescent shape then they really *are* in trouble of the direst kind, for this is a sure sign that they have passed right *out of the galaxy* and are now traveling in metagalactic space parallel to the galactic plane. If the signs continue they will know they are heading into the abyss between galaxies on a course that leads to nowhere.

There is really no solution to this problem, or at least none that we can presently suggest. Presumably a civilization so much ahead of our own will have come up with the necessary answer. It seems very improbable they would head out toward another star unless they had (suicide missions excepted!). Alien races that are going to travel with relativistic velocities have no other option.

One interesting and highly ingenious suggestion has recently been made, though in essence it is basically a form of dead reckoning.[4] This is to ignore completely what is going on outside the starship and instead re-create, utilizing a planetarium projector and a computer, an *artificial* chart of the heavens on a screen. The position of the ship being known accurately at the commencement of the journey, also its velocity and heading, it should be possible to upgrade continually the projected image as time goes by. It might thereby be possible to steer a course by dead reckoning, relying all the time on the computer to move star images across the screen, increasing or decreasing their brightnesses as they pass. But should the equipment prove faulty the star travelers might find themselves in the very unenviable position of not knowing their position.

In closing this chapter we must inevitably wonder how many strange alien vessels now ply endlessly through the galaxy, their occupants long dead, because in a realm of countless stars there was no star to guide them!

# PART II

# 8. THE CHEMISTRY OF ALIENS

In the ensuing chapters we will be considering and reviewing, so far as we can, the record of possible alien visits to our planet—past, present, and future. Inevitably, the mind of the reader will turn to the particular forms such beings might assume. First impressions, helped along by the endeavors of science fiction writers, may probably be that the question is wide open. It is certainly open. In the circumstances it could hardly be otherwise. It is doubtful, however, if it is quite as open as is popularly believed.

There is little doubt that science fiction has done much to foster the belief that creatures from other worlds will be not only different but extremely grotesque. Consider the words of one of the earliest and greatest of science fiction writers, H. G. Wells, as he describes the invading Martians in his renowned classic *The War of the Worlds.*

> They were, I now saw, the most unearthly creatures it is possible to conceive. They were huge round bodies—or rather, heads—about four feet in diameter, each body having in front of it a face. This face had no nostrils—indeed the Martians do not seem to have had any sense of smell—but it had a pair of very large, dark-coloured eyes, and just beneath this a kind of fleshy beak. In the back of this head or body—I scarcely know how to speak of it—was the single tight tympanic surface, since known to be anatomically an ear, though it must have been almost useless in our denser air. In a group round the mouth were sixteen slender, almost whip-like tentacles, arranged in two bunches of eight each.

The motion picture has escalated the "horror" aspect to a considerable degree. To some extent this is both forgivable and understandable. Films are intended primarily for purposes of entertainment and it is unlikely that script writers and producers intend their "brain children" to be taken seriously. When the average layman thinks of possible beings from other worlds he expects creatures that are different—different in his mind being synonymous with grotesque. He is rarely disappointed! Unfortunately, this leads to a totally erroneous concept of the possibilities. The belief is nurtured that aliens come in a variety of the most improbable forms, that they are almost certain to be hostile, and that this hostility will manifest itself in diverse horrible ways. By implication too is planted the impression that the laws of physics, chemistry, and biology (especially the last) must be quite different on other worlds. Such a premise is completely unacceptable. Alien beings will undoubtedly have to conform to prevailing conditions on their respective planets. They will be subject to differing evolutionary pressures and, as a consequence, must in many instances differ from ourselves to a greater or lesser degree. This does not mean that they must of necessity be either grotesque or repulsive. A cat or a dog is totally different in form and external characteristics from a man, yet neither of these creatures could justifiably be termed grotesque or repulsive—quite the reverse in fact. Admittedly, there may be worlds where environmental conditions have created and molded life-forms highly repugnant to our eyes. This does not automatically render such beings monsters. Even less does it imply minds that are evil. Beauty, it has been said, is in the eye of the beholder. So, no doubt, is the lack of it! These things should be remembered—especially if we were suddenly to be confronted by highly intelligent arthropods from a world of the star Epsilon Indi.

We must be prepared to accept that intelligent and cultured beings from other worlds may vary greatly in their external characteristics. Some will not be all that dissimilar from ourselves (perhaps even improved versions of the model!). Others could be quite different although not necessarily in a horrifying or repulsive way. Science fiction novels and films tend to present us with the two extremes. We have either pure "terrestrials," perfect in an almost godlike way—men of startling physique, women of superb shape and form—or monsters of the most repulsive sort. If we settle for a range of forms between these two we will probably be on fairly safe ground.

Bearing this in mind it might be both interesting and worthwhile to look at a few of the possibilities in the light of simple basic chemistry. It is not, however, our intention to go deeply into chemistry in a book of this sort but merely to scratch the surface. This will serve as a good background—perhaps even a good and necessary brake—in considering extraterrestrial visitations past, present, and future.

First, it is necessary to make certain basic assumptions. One of the most basic is that life will evolve to suit prevailing conditions and will subsequently adapt itself to any changes occurring in these conditions. This is true of terrestrial life, which originated in a reducing ammonia-methane primary atmosphere then gradually modified itself to the present oxygen-nitrogen system.

The second essential assumption is that life must have an inbuilt system containing information about itself and have also the necessary powers of retrieval to draw on these genetic "information banks" and subsequently translate the material into the necessary chemical enzyme responses. A study of the chemistry of DNA, RNA replication, enzyme and polypeptide (protein) formation provides abundant evidence of how well this works on our own world.

Genetic information storage and retrieval is vitally necessary, for without it mutation would lead to haphazard development of unsuited life-forms. Put more simply, the life form would "forget" and pursue a largely chance path of development. When we contemplate some of the science fiction and motion picture "aliens" we must assume that complete amnesia has overtaken the progenitors of the creatures portrayed! Nature can, of course, make mistakes. A case in point is probably that of the giant reptiles which were virtual lords of the Earth during the Triassic, Jurassic, and Cretaceous ages. These great and ponderous creatures appeared, lived, and disappeared all within a short period of time, relatively speaking. The environment changed but their genetic mechanism did not or could not adapt itself. They were quite simply one of nature's mistakes.

The assumptions so far adopted have dealt only with fundamental requirements and not with the capacity of specific elements to fulfill a biochemical role. On this planet it is carbon upon which life is based, but so far at least we are not assuming the absence of an appropriate alternative. Similarly, in recognizing oxygen as the gas necessary for respiration we are not ruling out the possibility of chlorine or fluorine in that role.

The other prime essential to terrestrial life is, of course, water. It is something which none of us can do long without and indeed it constitutes the major component of our bodies, a somewhat humbling thought. In recent years there have been suggestions that an analogous role might be played by ammonia in the bodies of certain alien beings. In fact, a humorous cartoon appeared in a magazine a few years back based on this idea. Against a background of terrestrial desert, scorching sun, and limitless sand a very peculiar-looking creature has left its "flying saucer." As it staggers along it clutches its throat and gasps, "Ammonia! Ammonia!"

The concept of ammonia in this role has been seen as improbable on the following grounds:

a. Liquid ammonia *boils* at the very low temperature of -33°C. Consequently a life-form based on the compound must of necessity be a "low-temperature" one.
b. Liquid ammonia *freezes* at -77°C. This means that ammonia will exist as a liquid only over a temperature range of 44 degrees.
c. Enzymes (a sort of biological catalyst which help biochemical actions along) are rendered inoperative in ammonia because of its high alkalinity.

Facts (a) and (b), however, are valid only at normal atmospheric pressure, i.e., the pressure at which we live on the surface of the Earth and which has been given the value of 1. It is a fundamental fact of physics that alteration in pressure changes both boiling points and freezing points. (Try boiling water on the summit of a high mountain where atmospheric pressure is much reduced!) *Increases* in pressure *raise* both boiling points and freezing points. At a pressure of 60 atmospheres the boiling point of ammonia is no longer -33°C but 98.3°C, an increase of 131.3 degrees. Similarly, the freezing point is raised by 44.3 degrees to -32.7°C. At this pressure, therefore, ammonia has a liquid range of 131 degrees. This contrasts with the 100-degree liquid range of water at an atmospheric pressure of 1. Admittedly, a pressure of 60 atmospheres would not be to our liking. On the other hand, a pressure of 1 might be equally unsatisfactory to certain aliens. The point we are making is that ammonia-based life need not necessarily be a low-temperature one so long as atmospheric pressure is sufficiently high.[1]

Point (c) is less easy to explain away without reference to somewhat more involved chemistry. Perhaps it will suffice if we say that this problem arises (or appears to arise) because we are thinking exclusively in terms of terrestrial conditions. Under those of a radically different environment it could well disappear.

When we say that ammonia-based life-systems are impossible we are simply allowing ourselves to be blinded by the essential conditions of *our* environment. On the other hand, though there are similarities and parallels between water and ammonia in their liquid states, the chemistries of the two are far from identical. Inevitably these differences would have a distinct bearing on the form and type of life which could utilize ammonia as we do water.

We cannot rule out the possibility that on certain remote planets flora and fauna depend on liquid ammonia just as their equivalents here on Earth depend on water. Nature is both ingenious and highly adaptable. There are, however, many apparent difficulties in the way of such life. We stress the word "apparent" because we are familiar with life only as *we* know it. It is unlikely, though, that visitors to our world would, if they originated on an "ammoniacal" planet, find our environment other than very hostile (and of course vice versa).

Another favorite for consideration is a life-form based on the very common element silicon. Terrestrial life, as we have already stated, is based on carbon. In many respects silicon is similar to carbon, notably in its ability to link up with other elements. This is perhaps a little unfortunate since it has led to the impression in some quarters that it could easily be substituted for carbon. Not so, and in fact a better theoretical case for non-carbon life can be made for certain other elements. Carbon atoms link up with each other to form long chains, known to chemists as polymers. These links or bonds are stable to a degree shown by no other element, which is almost certainly why life on Earth is based on carbon and probably that in other parts of the universe as well. Silicon also has this affinity to link with many other elements but to a much lesser degree.

It is essential that an element that is to constitute the basis of an organism should be able to form *stable* polymers. Carbon measures up to this requirement very easily. So also does silicon. As a matter of fact, the rocks of our planet are composed to very large extent of silicon-oxygen polymers. The silicon-to-oxygen bond is very stable—indeed, far more so than the carbon-to-carbon bond to which we largely owe our existence. Even the silicon-to-silicon bond has considerable strength though substantially less so than the carbon-to-carbon.[2] Why, then, is terrestrial life carbon-based and not silicon-based? The answer to this may lie in the great abundance of water on our planet in the presence of which any *complex* silicon compound is liable to be acted upon (or "hydrolyzed," as chemists would put it) to silicon dioxide ("silica"). In the circumstances water could hardly fulfill the role we mentioned a few paragraphs back. Water and silicon compounds of a complex nature do not make good companions. We may therefore assume that beings whose bodies are based on the element silicon would require to exist on worlds devoid of water. This rules out terrestrial-type planets. Even the very minute amounts of water believed to be present in the atmosphere of Mars could be detrimental to any such creatures. Silicon-based life is a theoretical possibility so long as water, ammonia, and compounds of a similar chemical nature are not around, for these render complex silicon polymers unstable. This greatly lowers the chances of finding silicon-based life in the universe. In most other respects, however, silicon can emulate carbon in the structure of the compounds it produces, especially with respect to those containing oxygen and nitrogen.

It is also difficult to conceive of a way in which silicon-based life could evolve from simple inorganic materials. Here the analogy with carbon does not help at all. We can postulate how organic chemical compounds based on carbon may have been formed from the primeval inorganic "broths" of the early terrestrial oceans. If we try to do the same with silicon we come up against problems of both a chemical and a biological nature. This does not completely rule out the possibility, for the problems probably arise from our relative ignorance of silicon chemistry at the present time. It is fair to point out that we are still very far from clear as to the precise manner by which *carbon-based* life was able to evolve from simple inorganic materials. The fact that it could—and did—is borne out amply by the existence of the writer presently typing these words and the reader eventually reading them! Perhaps, for all we know, at some remote point in the universe at the present time a "silicon" man is wrestling with the same problem with respect to carbon-based creatures! We wish him well.

Another possibility that must be considered is that group of highly reactive elements known to chemists as the halogens (fluorine, chlorine, bromine, and iodine). Of the four, bromine and iodine should probably be excluded since their relatively large atomic size seriously restricts the number of possible compounds they can form. It is also a fact that these two elements are comparatively rare in the universe as a whole.

Many stars show spectra in which the preponderance of chlorine and bromine exceeds that of oxygen. In instances where such stars possess planetary families we might anticipate halogen-rich worlds. These are hardly worlds that would appeal to our kind—oceans that are solutions of hydrochloric or hydrofluoric acids, atmospheres of chlorine or fluorine.

It is interesting to reflect on the manner in which respiration might proceed if living beings inhabit any of these worlds. A possibility so far as plants are concerned is as follows. During the hours of daylight terrestrial plants take up carbon dioxide from the atmosphere and, by the well-known process of photosynthesis, transform this into a compound of carbon and oxygen which we know as starch. At the same time plants give out oxygen into the atmosphere. The process is actively helped along by the green coloring matter in plants known as chlorophyll. An analogous mechanism is seen as a possibility on a "halogen" world in which plants would take up carbon tetrachloride from the atmosphere (a likely constituent of a chlorine atmosphere), by photosynthesis turn this into a carbon-chlorine compound (analogous to starch), and emit chlorine to the atmosphere under the influence of some kind of "halogen chlorophyll."

Chlorine is more likely to be involved than fluorine. The latter is an intensely reactive element (as anyone who has ever had to work with it will readily testify).

In theory at least there is nothing fundamentally wrong with the system we have just considered. Indeed, with regard to photosynthesis, it is considered by some biochemists as preferable since the reduction of carbon tetrachloride is seen as an easier and more efficient process than the analogous reduction of carbon dioxide.

Trying to postulate a respiration system for animals (intelligent and otherwise) is considerably less easy. Such creatures would probably inhale diluted chlorine just as we inhale diluted oxygen and exhale perhaps carbon tetrachloride as we exhale carbon dioxide. Oxygen is carried by the blood in our bodies to all the essential tissues by a compound known as hemoglobin. We might therefore envisage chlorine being carried to alien tissues by a suitable hemoglobin equivalent in their "blood." The main difficulty of this straight analogy is the very reactive and corrosive effect of chlorine. Flesh and body tissue able to withstand continued contact with this element are going to be very different from the terrestrial equivalents. As a guess we might suggest tissue would be dark and leathery. Creatures of such worlds, if they exist, are almost certainly going to be very different from us—perhaps alarmingly so. We can probably take comfort in the thought that such beings able to reach our world would be at a massive disadvantage and would be most unlikely to linger.

In comparing and contrasting biochemical systems with our own it is very important to remember that atoms of the various elements differ greatly in size and this in turn alters and influences their respective chemistries to a considerable degree. Compounds formed, though sometimes analogous and therefore similar in many respects, will also reflect this difference. Consequently, we must take into account what this will do to the highly involved biochemistry governing reproduction, growth, and other body processes.

Suppose, for example, we select a number of compounds or polymers essential to terrestrial life. These will be based principally on the elements carbon, oxygen, hydrogen, and nitrogen. On paper it is very easy to substitute one or more other elements and produce a chemically analogous compound. In some instances it is even possible to give this practical expression in the laboratory. The trouble can start when we try to fit the analogous chemical compound into the biochemical scheme of things.

Prevailing temperatures on a planet must also be expected to have some bearing on the type and form of life produced. It is reasonably certain that a proportion, perhaps even a high proportion, of the planets within our galaxy (as well as in others) will be colder, possibly much colder than our own. This is certainly true so far as our solar system is concerned. Mercury and Venus are both much hotter than the Earth but all the rest are considerably colder. We cannot say how typical our solar system is, but it seems reasonable to expect other solar systems to be roughly similar. Should a planet be too cold it would be unreasonable to anticipate the existence of life on its surface. There could, however, be many much colder than Earth but still not too cold for life to take hold and survive. Intelligent beings arriving here from worlds in this category would probably regard our polar regions as tropical and our tropics as unspeakable!

To exist in a low-temperature environment fairly reactive elements would appear to be necessary if the beings concerned are to think and function as we do.[3] This brings us to what we might term time scales, and whether other life-forms would opt for one similar to our own. This is a concept that requires a little amplification. A single thought (not necessarily an intelligent one!) occupies a few milliseconds in the case of our own kind on Earth. According to zoologists, quite a few of the giant prehistoric lizards found thought a very protracted business, which is merely a polite way of saying they were slow-witted! This probably had a considerable bearing on their eventual demise and total disappearance. Even the rate of our movements is intimately tied up with our chemical equilibrium and this can be slowed down under conditions of extreme cold. Polar explorers are well aware of the numbing effects of cold on both thought and action. In permanently frigid conditions some compensating mechanism would be required to enable our brains and bodies to function satisfactorily on the time scale which we regard as normal. But life on other planets might not conform to our time scale. We might go to extremes and envisage beings taking several hours just to wake up or some scurrying around like the characters in early speeded-up movies! Since we would expect life to develop a kinetic pattern in keeping with the environment in which it exists and which produced it, it seems much more reasonable to assume that a sensible and universal "Mother Nature" will have elected to work within finer limits. Nevertheless, some degree of difference in this area might be anticipated. On worlds having temperatures higher than those of Earth, reactions, physical as well as chemical, could be a little bit more rapid, and on worlds with lower temperatures the reverse would be happening. On our own world we are familiar with the fact that cold-blooded creatures are generally lethargic in low temperatures and active in high.

Variations in pressure could be as relevant as variations in temperature. Increase in pressure on a chemical reaction has the effect of bringing the reactants into closer and more intimate contact with one another, thereby increasing the rate of the reaction. In a biochemical context this can lead to an increase in the metabolic rate, unless some compensating capability (a kind of biological negative feedback) is built in.

*High atmospheric pressure* might be adequately compensated by the existence of *low temperatures* resulting in life-forms with a metabolic time scale more in keeping with our own. It is interesting to speculate on the probable effects of *high temperature* and *high pressure.* Metabolic reactions in such circumstances could be very rapid indeed and the poor creatures would probably be "burned out" in no time, a short life and an active one! Some form of enzyme activity having an inhibiting effect might serve as a compensatory biological mechanism.

Within our own solar system there are planets having high pressures and gravities as well as low temperatures. These are the "gas giants" (Jupiter, Saturn, Uranus, and Neptune). These are of more interest in a life context than was once supposed, though it would certainly be unwise to think in terms of alien cultures and civilizations. (Jovians, Saturnians, and the like, we can be sure, belong exclusively to the pages of science fiction.) Conditions on such planets might not be inimical to the *beginnings* of life. Those presently prevailing on Jupiter are not known with any degree of certainty. It is thought there could be three regions or layers which gradually merge into one another: an aqueous layer (cloud), a high-pressure region, and an "ocean" of liquid hydrogen and some of its compounds.[4] In these circumstances there would thus be a region appropriate to the initiation of life employing water as its basic solvent material, one in which ammonia would fulfill an analogous role, as well as one (the "hydrogen ocean") containing so many different possible starting materials that almost anything could happen. A similar state of affairs might conceivably prevail on the other three "gas giants."

We must here return to a point raised briefly a few pages back—the uniqueness or otherwise of our solar system. Now, however, we are thinking more in terms of the chemical elements present and their relative proportions to one another, and in this area we have reason to believe the solar system to be typical on two counts.

1. The composition of space is such that condensing stars and planets are likely to have a composition and distribution similar to that of the solar system.
2. Though the planets of the solar system can be roughly grouped into two basic types (solid bodies and gas giants) it cannot be denied that no two of these planets are alike. It appears that on formation of a planetary system a kind of cosmic "fractional distillation" takes place, that is, there is a separation of the constituents.

We believe that our sun and solar system condensed from the same gas cloud more or less around the same time, give or take a few millennia. Elements present in the original gas cloud should therefore now be present in the solar system. Thus if we examine the spectra of other gas clouds we should be able to predict the constituents of planetary systems which may ultimately condense from them.[5,6] On the whole the proportions of constituents are akin to those of the solar system. There are, nevertheless, exceptions and these are of considerable interest. Certain stars are known in which the elements manganese, mercury, silicon, and calcium are at a higher concentration than in the solar system whereas their oxygen concentrations are much lower.[7] Such stars must have condensed from gas clouds showing the same constituent ratios. If one of these stars, in the fullness of time explodes as a nova the residual dust and gas cloud, which is all that will remain of the star and any planets it happens to possess, will be rich in these elements. Eventually other gas clouds will form and condense to stars and planetary systems. If they occupy this same region of space much of these elements will be available. Another feature of novae is their capacity to synthesize other elements.[8] In other words, some regions of space must have heavy concentrations of less usual elements. The implications with respect to equally less usual life-forms are therefore obvious.

What conclusions can we draw? It is probably a sanguine person who would attempt to draw any. It would seem, however, that since space on the whole is a fairly homogeneous medium, life-systems will for the most part have a restricted choice of elemental "building blocks." The major constituent available is undoubtedly hydrogen, followed by helium, carbon, nitrogen, oxygen, fluorine, neon, sulfur, chlorine, and argon.[9, 10] Helium, neon, and argon are all inert gases, almost totally unreactive and therefore invalid in a biochemical context. Hydrogen, carbon, nitrogen, and oxygen are the most freely available—and significantly it is on these that terrestrial life is based. Life-forms using ammonia in place of water could equally well use these (ammonia being a compound of nitrogen and hydrogen, whereas water is one of oxygen and hydrogen). Halogen-breathing life would require chlorine, fluorine, hydrogen, and carbon. All are present in quantity. It should be noted, however, that silicon is missing from the list of "plentifuls," though it *is* one of the elements known to result from certain novae. This tends to downgrade the possibility of silicon-based life.

Just how viable are biochemistries other than our own likely to be? To this question there can be no categorical answer. For a long time there was a strong tendency to dismiss any alternatives but now a much less rigid outlook is becoming perceptible. In the past many have been willing to accept the existence of alien life but only of a kind biochemically akin to our own. This has led not infrequently to the most incredibly bizarre extrapolations involving carbon/oxygen/hydrogen/nitrogen (i.e., terrestrial) life, where in the postulated environment recourse to, say, a chlorine-based system would probably have been simpler. Small wonder at times that science fiction has treated us to so many BEMs (bug-eyed monsters). It is admittedly difficult to envisage metabolisms of which we cannot have knowledge, but neither can we deny a possibility simply because we have no experience of it. There has to be a middle road.

Visitation of our planet by aliens implies not just the existence of life elsewhere in the universe but of intelligent, technologically advanced life. A world could be densely populated by hordes of "creepie-crawlies" living in a luxuriant jungle. In all respects this would be a world teeming with life. It is not a life, however, that is likely to get very far in space. Certainly it is not going to reach Earth unless the passage of time and the progress of evolution bring an amazing degree of change. Such evolutionary progress does not necessarily imply enhanced intelligence.

The phrase "intelligent life-forms" is one that finds considerable usage in the jargon of exo-biology. Earth is said to be peopled by intelligent, thinking, reasoning beings. If we remove man from the picture, then it is no longer true—or only in a most limited sense. Certainly beasts have a greater degree of intelligence (or is it built-in instinct?) than primitive life-forms, but whether or not this is indicative of gradually evolving intelligence is difficult to say. On present evidence, however, it seems most improbable that Earth will ever be ruled by a race of intelligent cats and dogs. Pressure of survival caused beasts to develop fang and claw, the capacity in many cases for very fast movement, and the ability to climb. None of these things happened to man. He merely grew a greater brain and thereby developed skills with weapons and implements. In the course of time these have ranged from bows and arrows to intercontinental ballistic missiles, from crude plows to combine harvesters.

We must therefore ponder on whether or not the existence of developed life on remote planets automatically means the development of high intelligence. We see that this is what took place on our own world though only really with respect to one species, our own. Once again there can be no definite answer. Yet surely what happened on the third planet of a very average star two-thirds of the way from the center of a quite average galaxy cannot be unique. Reason refutes the idea. This may not be strict logic, but it does seem like honest-to-goodness common sense.

And now to the crux of the entire business: Would alien beings likely to reach us from the stars resemble us in any way? Would such visitors be sufficiently manlike for us to feel some common bond with them? We have already dwelt on possible alternative chemistries, pondered their translation into plausible biochemistries, and considered briefly how varying planetary constitutions and physical parameters could affect their metabolisms.

In science fiction there has long been a tendency to depict a very different life-form as malignant and evil and one in a mold like our own as a paragon. This is probably just our own conceit and arrogance showing through. Common sense ought to tell us that this represents merely a highly idealized state of affairs. It cannot be denied that a starship from another solar system landing near one of our major cities tomorrow would be very much more welcome were it to contain men and women like ourselves (the erect, bronzed, fair-haired gods and goddesses of science fiction?). Were it instead to deliver a load of highly intelligent and sophisticated arthropods, the reaction is easily predictable—even though these "spidery" visitors happened to be representatives of the kindliest and most amenable race in the entire universe.

Visitors from space are not going to be facsimiles of ourselves unless the worlds from which they come bear more than a superficial resemblance to our own. On a planet possessing a deep, thick atmosphere, life, assuming it can exist at all, must out of sheer necessity adopt a radically different form. The lower layers of such atmospheres could be very hot. Beings in environments of this nature might resemble floating balloons, feeding on organic compounds produced as a result of electrical discharges in the thick atmosphere.

Giant-insect protagonists maintain that some alien environments may have led to unrestricted insect development, which brings us back to the arthropods! No one can disprove this. The fact that such a thing did not happen here might be only fortuitous. However, our knowledge of insects gives us good reason for believing this to be an unlikely possibility. Aliens in the form of giant insects frequently make their appearance in lurid science-fiction films. On Earth insects abound, but despite their numbers they have never become dominant. Elsewhere in the universe the positions might have become reversed. In time they develop a supra-technological society, attain deep-space capabilities, and eventually descend upon Earth. Those in the science fiction films as often as not closely resemble terrestrial insects except of course that they are very much larger. Herein lies the fallacy. There are a number of reasons, all very sound ones, why insects cannot attain huge dimensions.

It is a fundamental and indeed obvious fact that all creatures must respire; that is, they require a supply of oxygen delivered to the essential tissues of their bodies. In most fish this is achieved by means of gills, a very efficient mechanism since air dissolves only to a very limited extent in water. Mammals have lungs. This is a complex system which has evolved in creatures whose ancestors left the water for dry land. It too is highly efficient but rather less so than gills—for the good reason it doesn't have to be. Oxygen in air is abundant. Insects, however, do not have lungs and neither for the most part do they have any gill system. In their case air is "piped" almost directly into close contact with the body tissues consuming oxygen by what is known as a tracheal system. Such a system opens to the exterior of the insect by spiracles (narrow tubes) of which there are generally ten pairs, two on the thorax and eight on the abdomen. These tend to be fairly complex structures because of four essential requirements: (1) prevention of foreign body entry, (2) prevention of water entry, (3) limitation of water loss while permitting ventilation, and (4) allowing "moulting" to take place (the casting of the external skeleton and subsequent development of a new one).

The tracheal system develops into a number of very small branches ending in what are generally termed tracheal end-cells. From each of these a number of extremely slender tracheoles extend as intra-cellular tubes reaching into various organs of the body. Air will not automatically pass through this complex, tortuous path. If the tissues of the insect are to be supplied with oxygen that oxygen not only must be drawn down into the system, it must also be enabled to enter the *blood* system. The latter involves transfer ("gas exchange" is the usual term) across the tracheoles or "between-cell" tubes. This takes place across the walls of these since they fulfill the essential conditions of being both wet and water-permeable. In most biological membranes water and oxygen permeability go hand in hand so this seems a reasonably likely region for oxygen transfer. The tracheae themselves (the main, larger-diameter tubes) are dry and non-wettable but have very thin walls and it is here that some biologists think oxygen transfer to the blood takes place. This is an interesting question, but one as yet unresolved. Respiration across a dry, non-wettable surface would be distinctly advantageous since it would involve absorption of oxygen by the body, coupled to minimum water loss.

Whatever the precise mechanism of oxygen transfer to the blood, air must certainly be drawn into the tracheal system. In other words, there is the essential requirement of active ventilation. The form this takes is dependent on two factors: the length of the tubes (directly proportional to the size of the insect) and rate of oxygen consumption by the body tissues. Small, active insects such as flies are able to manage by the simple expedient of oxygen diffusion through the spiracles and the tracheal system. Certain large *inactive* insects do the same—for example, large caterpillars. On the other hand, insects such as locusts maintain a *flow* of air through the spiracles, air sacs, and primary tracheae by virtue of muscular contraction. In effect this is really a kind of pumping mechanism brought about by rhythmic squeezing of the air sacs in conjunction with accurately timed opening and closing of the spiracles. As a mode of respiration this is outstandingly effective, since oxygen consumption by the tissues of some insects, especially their flight muscles, is extremely high. Such consumption is indicative of a tissue metabolic rate far in excess of anything achieved in mammals. Despite such exacting requirements the tracheal system is able to deliver the necessary oxygen. Why then, it might reasonably be asked, have not mammals adopted such a system? The answer is not hard to find. Diffusion is adequate only over *very short* distances. Since tracheal tubes in the case of a large creature would require to be disproportionately long, this would necessitate provision of some kind of efficient pumping system which would in itself increase the requirement for oxygen. Here, then, we have one of the two main factors inhibiting the size of insects. The other is the limitation imposed by the external skeleton, for as the insect grows larger this must become disproportionately large. For these facts we ought perhaps to be thankful. Because of their efficiency and numbers, insects, had they been able to attain large dimensions and proportionate physical strength, might by now have dominated this planet.

These are the reasons why giant insect-type beings from the stars are unlikely. It would be as well to bear them in mind when next the wide screen portrays a giant spider creature from Antares or Vega carrying off a terrestrial heroine. She shouldn't worry too much—such a creature can't exist!

Screen and story also at times treat us to fishlike creatures. There will, we must suppose, be wholly water-covered planets. Is an ichthyological society possible? Fish, as we well know, cannot live out of water. Gills, though highly efficient at absorbing the relatively small amounts of oxygen dissolved in water, cannot cope with air despite its much greater oxygen content. The reason for this is quite simple. In water, which has a high density compared to that of air, the gills are *supported* by the medium. As a direct consequence their very extensive surface area is exposed to the oxygen-carrying water from which oxygen is drawn into the gill blood vessels, carried to the heart, and pumped around the body to the various organs and tissues by means of arteries. In air, however, the gills clump or bunch together so that no appreciable surface area is exposed for oxygen absorption. Consequently the fish suffocates.

Thus, if an ichthyological civilization ever aspires to interstellar travel (most improbable in the circumstances of a water environment we would think) and reaches the land surface of Earth, then their ship must almost certainly be a giant water tank, a space aquarium in which they could be only observers at best, prisoners at worst. If they looked at all menacing there would be an obvious, quick and easy answer!

However, circumstances might not just be so straightforward. There are a few terrestrial fish which do not rely entirely on gills for respiration. One or two of these types can spend limited periods out of the water. There is, for example, the mudskipper Periophthalmus. This is a goby which inhabits tidal marshes. It is claimed that some species of this genus have greatly reduced gills and are compelled to *breathe air*. Another mouth-breather is the common eel Anguilla. Neither can we overlook the catfish Clarios, which inhabits water deficient in oxygen and has an efficient gill system, so efficient in fact that to a limited extent it can cope with *atmospheric* oxygen. Consider too the renowned electric eel, which has a large number of projections or papillae intruding into its mouth cavity. These contain small blood vessels (capillaries) the blood supply of which is linked to that of the gill system. The gills themselves are reduced in extent and in fact the creature drowns if it cannot reach the surface of the water from time to time in order to gulp air.

Certain fish are capable of swallowing air bubbles and absorbing the oxygen from them through the blood vessels of the gut. Others have a crude form of lung, notably the Neoceratadus of Australia, the Lepidosiren of South America, and the Protopterus of Africa. The first of these *must* use its lungs as well as its gills in order to survive. The other two are mainly air-breathing and *drown* if kept from the surface.

Could adaptation eventually go farther than this? Who can tell? Could it have done so already in worlds far from our solar system? Again, we cannot say, but it probably has. Whether this could in time lead to a race of fishlike intelligent creatures is a much more debatable point. If beings are to develop in the sense of creating societies it seems a lot more reasonable to assume that they will do more or less what they did here—gradually evolve into true land creatures, totally air-breathing with lungs. Locomotion for fish that can breathe air and survive on dry land represents something of a problem. Presumably the process of evolution will eventually take care of that too. It obviously did so on Earth, although it was a long time doing it. Until it was complete *we* were not able to go to the moon. If we take all these factors into account it seems unlikely we will ever be confronted with fishlike aliens.

It is difficult to escape the conclusion that alien races must tend more toward the mammalian type than any other. By this of course is meant intelligent races having high technological abilities. Nature prefers a logical answer to her developmental problems. Intelligent arthropods and calculating fish just do not seem to constitute such an answer. Our visitors from the stars could, nevertheless, be erect, biped hominids essentially the same as ourselves and still be starkly and frighteningly different. In some respects this prospect might even be worse—to be confronted by strange, alarming caricatures of ourselves.

Table 1 outlines some of the more likely possibilities. It must be emphasized, however, that these are only reasoned guesses—all that is feasible in the circumstances. There can be no question of drawing up rigid categories. We must envisage a range of conditions in which the terrestrial type gradually merges into one or other of the alternative types and vice versa. Nor is this all. The almost infinite number of possible permutations must be considered.

For example, a planet might be of terrestrial type with respect to dimensions, gravity, and thickness of atmosphere but that atmosphere might be other than the oxygen/nitrogen mixture which we call air. Result: beings perhaps humanlike in many respects but with a leathery, scaly skin depending on the nature of the atmosphere. The converse is also possible: squat beings with thick limbs and powerful muscles on a high-gravity, thick-atmosphere planet. Were that atmosphere of air, the respiratory system of such creatures could be one closely akin to or identical with our own.

TABLE 1

| *Type of Planet* | *Type of Life* |
|---|---|
| Terrestrial. | Similar to ourselves in most respects though in many instances liable to be something of a caricature. |
| Light-gravity; thin, oxygen-deficient atmosphere. | Tall; perhaps fragile structure compared to humans; weak muscles; large lung capacity, perhaps additional respiratory organs (e.g., external "air gills"). |
| Heavy-gravity; thick, erosive atmosphere. | Squat, thick limbs; powerful muscles, dark leathery skins or scale. |

There are a few less obvious possibilities worthy of a brief mention. Visitors to our world from one whose parent sun is a source of intense ultraviolet light would probably find themselves at a very considerable disadvantage here if nature as a consequence had rendered their eyes responsive only to ultraviolet light. How would they cope during nighttime on their own planet? Precisely as we do no doubt, save that the artificial lighting they employed would be in the ultraviolet region of the spectrum. Presumably they would require to use such facilities here both night *and* day.

Imagine instead that our visitors were from the world of an infrared sun. Unless their eyes were heavily protected they would find conditions here very trying. On the other hand, they would be at a considerable advantage during the hours of darkness since presumably each and every one of us would be visible to them by virtue of the heat radiation which our bodies emit.

The possibilities are almost limitless. Perhaps the only real certainty is that we will never know what to expect until they arrive—and then we will not know what to do!

# 9. UNTOUCHED BY HUMAN HAND

It may be appropriate at this point to make a more analytical examination of the possibilities of aliens reaching this planet. Though our theme is that of visits past, present, and future, we will restrict ourselves for the moment to the mathematical probabilities of visits to the solar system by starships in the past.

We believe the solar system to have been created between 4,500 and 5,000 million years ago. In common with all the other stars in our galaxy the sun has a motion of its own. During this immense period of time it is believed to have made about twenty circuits of the galaxy. This score or so of "galactic orbits" means that the sun has wandered extensively among the other stars and while so doing will have had many thousands of different stars as its near neighbors. At the present time the near neighbors of our sun are stars such as Alpha Centauri, Barnard's Star, and Sirius, to name but three. These have not always been our "close" companions and, given long enough, they will part company with us to be replaced by others.

Just how many different neighbors a star will have in its journeyings will depend on the position it occupies within the galaxy, its proper motion, and the density of stars within particular localities. How does all this affect the sun? At present, 10 stars lie within 10 light-years of us. A million years ago none of these particular stars lay within this distance but others did. Whether the number then was 10, 9, 11, or 12 we do not know, but we can be fairly sure that it was something of this order. In other words, the number of stars within 10 light-years of the sun has remained fairly constant.

If next we consider a larger volume of space around the sun, say, for example, 25 light-years, the number of neighbor stars will be increased to something like 140. This is only another way of saying that as the volume of space increases so also does the number of stars, which is what we would expect.

During the last 2,000 million years some 96,000 stars must have approached to within 10 light-years, 540,000 to within 25 light-years, and 9 *million* to within 100 light-years. Further extrapolations become pointless since they involve increasingly large distances. At least they are pointless if galactic exploration and colonization by orthodox methods of star travel are invoked. If, of course, bizarre sophisticated techniques are possible, such as those we considered in Part I, then distance might be almost meaningless. For the present, however, we will think only in terms of advanced orthodox technology.

On the scale of stellar distance 10 light-years is fairly insignificant. However, 96,000 stars entering or leaving a sphere of this radius is far from insignificant at a first glance. It becomes negligible when we realize that this represents only about *one* every 20,000 years! Using conventional techniques 10 light-years is also a very vast distance–60 million, million miles. It has been calculated, however, that about once in 11 million years a star may pass within ¾ light-year of the sun (about 4½ million, million miles).

This is the general background against which we must assess the chances of alien visitations by more or less orthodox techniques. We should probably be cautious regarding our use of the word "orthodox" in this context. Here orthodox simply implies techniques that do not invoke the type of instantaneous or near-instantaneous transit that travel through "non-space" or a Black Hole might imply.

The two main difficulties in the assessment we are trying to make are the fairly obvious ones.

1. To what extent has intelligent life proliferated within the galaxy?
2. To what extent has such life achieved interstellar capability and what is the potential of that capability?

The Milky Way, the great galaxy to which we belong, is almost certainly very much older than the solar system. Recent estimates put its age at something like 13,000 million years, which is about two to three times older than our sun and its family of planets. There has, therefore, been ample time for life to be initiated, develop, and become highly and technologically sophisticated, and to achieve this even before our sun and solar system were created. We know too that there must certainly have been sufficient stars of the right kind. Thus, by galactic standards, we may be a very primitive race indeed.

A glance at a photograph of any galaxy shows that star density varies considerably from region to region. There must have been periods during which more stars were close neighbors to our sun than others. Although no allowance has been made for this in the relevant part of Table 2, the figures given are likely to represent reasonable averages since there must also have been periods during which the sun lay in regions of *low* stellar density.

TABLE 2 [1]

| Distance from sun (light-years) | 10 | 25 | 100 |
|---|---|---|---|
| Average number of neighbor stars | 10 | 140 | 9,000 |
| Average number moving "in" or "out" of this region/$10^6$ years * | 48 | 270 | 4,500 |
| Average time remaining in region (years) | 210,000 | 520,000 | 21,000,000 |
| Entered or left during last 2,000 million years | 96,000 | 540,000 | 9,000,000 |

* $10^6$ = 1 million

To what extent will life have proliferated in our galaxy? This is a fundamental question but by no means an easy one. Estimates over the years have been many and varied—probably because of the sheer lack of reliable guidelines. Equally many and varied are the estimates of how long alien societies (and presumably also our own) can be expected to exist. Here too reliable criteria are hard to find. Even the estimated number of planets likely to represent suitable habitats for the initiation and development of advanced technological cultures varies between 200,000[2] and 10,000 million.[3]

Since we are considering the number of stars entering or leaving particular volumes of space surrounding our sun over the last 2,000 million years and we accept our galaxy to be at least 13,000 million years old, we can divide these estimates by six. The number of appropriate planets then ranges from 34,000 to 1,700 million. Should we consider the lowest or the highest limit? In arguments of this sort it is probably always wise in the first instance to accept the lowest. If we are wrong, then the case we are trying to make is enhanced, not diminished. We will therefore initially take the figure of 34,000 but round it to 40,000. On this basis we will endeavor to calculate the chances of alien starships having come our way. Certain fundamental assumptions are called for if we wish to prevent our analysis from becoming unduly complicated. These are fourfold and are as follows:

1. The 40,000 advanced alien cultures are fairly evenly distributed throughout the galaxy with respect to appropriate stars.
2. They possessed star travel capabilities 2,000 million years ago.
3. No significant degree of liaison has taken place between them. We are thinking in terms of *separate* visits by *separate* races, not combined operations.
4. The life-span of these races is of indefinite length. We will ignore the effect which contact with lesser beings (people like ourselves) might have had on them since this would presumably be of negligible consequence to advanced galactic communities.

Forty thousand stars (which there must be if there are to be 40,000 advanced societies) may seem a fairly considerable total. Nevertheless, it represents only about 0.00004 percent of the stars in the Milky Way. (The total number of stars in the Milky Way is considered to be $10^{11}$, i.e., 100 thousand million.) On this basis only one star in 2.5 million would possess such a society.

Earlier we saw that the likely number of stars coming within 10 light-years of the sun during the last 2,000 million years was 96,000; 0.00004 percent of this figure is approximately 0.04. Thus, there has been a 0.04 probability of a single visitation to the solar system by an alien expedition of 10-light-year capability during the last 2,000 million years. If on the other hand we accept the upper limit of 1,700 million suitable planets, the chances are increased 50,000-fold, which gives a probability rating of about 2,000 visitations by alien expeditions.

The disparity resulting from the high and the low estimates is so marked that a compromise figure is clearly called for. In selecting this it is preferable to err on the conservative side. The range 40,000 to 1,700 million is an extremely large one. A figure of 10 million appropriate planets seems reasonable. While considerably in excess of 40,000, it is nevertheless far removed from 1,700 million. On this basis the probability is that in over 2,000 million years there may have been visitations by alien expeditions on approximately ten occasions—that is, by alien expeditions having a 10-light-year capability.

A factor of considerable importance is the length of time for which an alien civilization will exist. Two thousand million years may be much too long. Again it may not. We have no reasonable criteria to judge by. Earth is still too young, apart from which there is no certainty that terrestrial standards would be universal. What too is the position regarding evolutionary change over such a long period of time? We cannot predict what will happen to our own form eventually. We may grow tails again or develop larger brains! Future biology on Earth is an enigma—which is perhaps just as well. Equally problematical is the extent to which a starfaring civilization will extend its cosmic frontiers. It does, however, seem more likely that the tendency will be for these frontiers to expand, for an even greater deep-space capability to develop with the passage of time, assuming of course that the civilization concerned is a sane one. As with us, their choice could be deep space—and a future; or deep-space, nuclear holocaust—and destruction! There are, too, all the other perils and follies—

pollution, world starvation, gross and frivolous waste of non-renewable metals, fuels, and other natural resources. About these *we* should certainly know!

We will suppose, however, that galactic communities are both long surviving and nonsuicidal and that having once reached the nearer stars their stellar frontiers continue to widen. Restraints imposed by distance will be temporary and due more to resource availability than technical inability. To a people of great maturity a 10-light-year capability is probably a gross underestimate, almost an insult.

Let us think now in terms of an interstellar capability of 25 light-years. Once again, the possible visitation rate is going to be a function of the proliferation of such peoples in the galaxy. A sphere of radius 25 light-years around our solar system means a greater number of starfaring civilizations having the ability to reach us.

Table 2 shows that during the past 2,000 million years some 540,000 stars may have passed the sun within a distance of 25 light-years. Calculating as before, this could mean something like 50 visitations over the period. If we think of a capability of 50 light-years there is a probability of over 200 visitations. Extending this to 100 light-years, visitation probability rises to between 900 and 1,000.

During the past two decades space vehicles and probes have become increasingly numerous in the solar system. They have reached the moon, landed on Mars and Venus, bypassed Jupiter. One has even headed out into the interstellar wastes. We on Earth are, quite rightly, very proud of these achievements. We readily believe that for the first time since the solar system was created artificial bodies have appeared among its worlds, moons, and asteroids. It may be, however, that the sun has shone on many occasions on metal objects, on peculiar vessels, while men of Earth still peered out from their caves. To certain galactic peoples our sun and its worlds could be a fairly familiar region. It might be wise therefore not to dismiss *all* UFO and flying saucer stories out of hand.

Our theorizing has been a little idealized. We remain uncertain about the length of time civilizations can survive and we have no clear guarantee that such civilizations will grant a continuing commitment to galactic exploration, although diminishing resources and expanding populations may well render it mandatory. Neither can we forecast the chances of such peoples meeting and the possible effects of any resulting liaison between them. We have no empirical data. We can only assess on the lines we already have but at least this is preferable to mere speculation.

If the solar system has been something of a stamping ground for aliens at intervals over the past two million millennia it is only natural to inquire whether or not any tangible evidence may remain in the shape of artifacts of one sort or another.

Perhaps we should think first about the debris so far left behind by our own, as yet, brief excursions into space. Already we have littered the surface of the moon to no small extent. Today its scarred and rugged surface bears the mangled remains of probes that were meant to hard-land and did, probes that were meant to soft-land and didn't, plus Apollo lunar modules sent crashing back to the moon when their purpose was fulfilled. There are also a number of items in reasonable shape: "moon buggies," remotely controlled devices, Apollo lunar module "legs," and a variety of other small expendable items. In an endless journey around the sun go satellites and satellite launchers. Around the Earth go many other devices, some of which may continue to do so for very long periods.

As space exploration from our own planet becomes more efficient and sophisticated, there will probably be less of this waste. Much of it to date has been unavoidable—for example the Apollo lunar capsules and the "moon buggies." There was simply no means of recovering these items, so for them it was a one-way trip. This could equally well apply to alien expeditions reaching the solar system. In space travels within their own planetary systems all such primitive and wasteful techniques would long have been abandoned and forgotten. But such races reaching our system, and at the limit of their range, might easily find that they must, for a time at any rate, revert to such practices. Used items and certain other equipment regarded as expendable and a cumbersome hindrance on the return journey could easily have been abandoned. If so, then in parts of the solar system they may *still exist.* We might have to be careful regarding wrong conclusions here. Suppose the first manned Earth expedition to Mars were to find some elaborate-looking piece of apparatus, or even the remains of an old boot. This does not automatically mean Martians. It *could* mean Alpha Centaurians, Epsilon Eridanians, Tau Cetans, or Capellans!

We can envisage alien visitors to our system in ages past dumping material either on planetary surfaces or in closed orbits. Such jettisoning could have been of fuels, spent reactors, and expendable equipment. But they might also have elected to leave behind items designed for other purposes, items intended to be found at a later date, items that might, assuming their finders could decipher the characters inscribed upon them, tell of the visit and of the race that made it. Additionally there could be items left behind relating specifically to certain aims of the expedition—e.g., sensor devices to record planetary conditions, relay and telemetric equipment to transmit such information, marker beacons for the benefit of future expeditions.

Since this chapter deals specifically with artifacts it is appropriate that we examine these possibilities in a little more detail. Artifacts left within the solar system by alien expeditions in the past could be grouped into the following categories:

*Space Laboratories:* Automatically controlled. Now probably in derelict condition unless installed within the solar system during the last century or so.

*Radio Relay Stations:* Purpose—transmission, reception, interception, and retransmission of signals from interstellar distances; part of a galactic communications network. Automatic and designed to function for extensive periods.[4]

*Telemetry Stations:* Stations designed to detect environmental characteristics and to record and relay the results.

*Marker Beacons:* Intended to transmit signals to which future starships from the same civilization might "home in" or use as navigational aids. Alternatively they might indicate known sources of useful fuels and mineral deposits or hidden dumps of essential equipment.

*Monuments and Edifices:* Nonfunctional obelisks or the like of historical significance indicating where and when a particular arrival occurred and the source of the expedition making it. Might also serve to mark the graves of dead astronauts.

*Implements:* Equipment ranging from lost screwdrivers to abandoned unserviceable computers and atomic reactors.

*Refuse:* Scrap and waste material of various types, metal, plastic, ceramic. Might also include chemically contaminated and radioactive material or biological waste.

*Environmental Evidence:* Fused rock, unaccountable radioactive "hot spots," paleomagnetic anomalies.

When we examine the possibilities as closely as this it would seem that finding evidence of alien visitations should not be a particularly difficult task. Some writers who have devoted whole books to the theme would appear to endorse this viewpoint. Unfortunately, the position is by no means so straightforward. Were it so we might by now have whole museums devoted to cosmic artifacts.

Before we begin our search for such artifacts, or impute cosmic origins to items of an enigmatic nature, it may be as well to consider how alien expeditions might have gone about their business. If such expeditions were few in number or landfalls on our planet only desultory, then incidence of artifacts will be rare. Moreover, if all or nearly all landfalls were carried out in early geological ages, then lithologic, tectonic, and geophysical changes will have ensured that these were either destroyed or so encapsulated and at such depths as to render their retrieval virtually impossible.

There are three general regions in the solar system in which we might expect alien artifacts.

1. On or below the surfaces of planets (certain ones), satellites, and large asteroids.
2. In orbit around planets, satellites, and large asteroids.
3. In orbit around the sun. Such orbits might be heliocentric (the sun at the center of the orbit) or hyperbolic (the sun not at the center of orbit after the fashion of a comet). (Figure 22)

Let us first consider the surfaces (or subsurfaces) of planets and other bodies. Right away we can eliminate the giant planets Jupiter, Saturn, Uranus, and Neptune. These are gas giants with probably no solid surface. Even if there were a solid surface the gravitational pull is so strong that no terrestrial mission could hope to survive. Any alien objects that spiraled down toward these planets have most assuredly gone for all time. This leaves us with

Mercury, Venus, Earth, Mars, Pluto, some of the satellites of the gas giant planets, perhaps such asteroids as Juno, Ceres, Pallas, or Vesta, and of course our own relatively accessible moon.

Planets with extensive atmospheres, and this really means Venus and Earth, are not particularly good prospects due to the erosion brought about by such atmospheres. The main hope with these is that any artifacts will have been buried but not too deeply. Even in these circumstances alkaline or acid corrosion could eventually have removed all traces. Much would obviously depend on the time and place of burial as well as the nature of the terrain. Volcanic, magmatic, or tectonic action, which in the case of Earth has been widespread, must also have taken its toll of any artifacts deposited on our planet. To some extent this constitutes something of a case for supporters of the idea of past extraterrestrial visitations when asked to explain the absence of physical evidence. Unfortunately it is rather a negative and unsatisfactory form of argument. Artifacts there may be on Earth, some *may* already have been found, but, generally speaking, the prospects on our own world are not very favorable.

The moon is a much better bet. Here there is no atmosphere to erode and degrade and though we have only a vague idea of the moon's early geological history, it has probably been stable for a very much longer period than the Earth. What then, if anything, lies beneath the dust and cinder of the lunar maria? Opponents may come forward with a seeming strong case. Six manned expeditions have so far reached the moon. Twelve men have walked upon its surface. Nothing was found.

This of course is no real argument. To date the area of the moon's surface trodden by man represents the smallest fraction of its total. Even those small areas have by no stretch of the imagination been thoroughly explored. For all we know, much of interest might be hidden under the very dust that now bears the imprint of human footsteps. And what of the "hidden" side, the side permanently turned from the gaze of Earth? What might have transpired in these regions in the long years before the Russian probe Lunik 3 made its first and now almost legendary journey around our satellite in September 1959? The moon has yet many secrets to yield—not all of them of necessity lunar.

We are thus left with Mercury, Mars, the asteroids, Pluto, and the satellites of the large "gas" planets. The position so far as Mars is concerned is probably midway between that of Earth and the moon. The atmosphere is a thin one largely of $CO_2$ and this, it is thought, is bound to have had a detrimental effect on any artifacts on its surface. We must also take into account the "sandblasting" which would almost certainly occur as a result of the tremendous and extensive dust storms that are so essential a feature of Mars. On the other hand, these could well bury and therefore help preserve artifacts, but only, perhaps, until another storm blew the sand away to some other part of the planet.

So far as the satellites of the giant planets are concerned we know that Titan and probably also Triton and Ganymede possess atmospheres. These are very tenuous and may not be too severe in their effect on any artifacts. Of surface conditions on the four principal asteroids (Ceres, Pallas, Juno, and Vesta) we know little, but certainly they have no atmospheres. The same can be said of our outermost planet, Pluto. Unfortunately, so far as we are concerned at the present time, all these bodies are too remote for any possibility of examination. Additionally, the principal asteroids and Pluto represent only a very small proportion of the total solid planetary surface within the solar system (about 17 percent).

This leaves us with little Mercury, closest planet to the sun and subjected to the most withering heat. How would artifacts fare on that harsh, barren surface? Depending on the metal of which they were made, some could soften, perhaps even fuse. Prospects here are not particularly promising, but they should not be ruled out.

An expedition reaching the solar system from a world of another star would in all likelihood devote the bulk of its time and attention to those members of the system closest to the sun—Mercury, Venus, Earth, and Mars. Certainly it is here they would start if their prime purpose was to establish the presence of life or of likely conditions for its eventual initiation and development.

What now of objects or devices existing in orbits around the sun or around the planets and their satellites? Any in hyperbolic orbits around the sun would be extremely difficult to detect and probably impossible to examine. So poor are the prospects here that they are best ignored.

Artifacts in closed orbits around the sun or in closed orbits

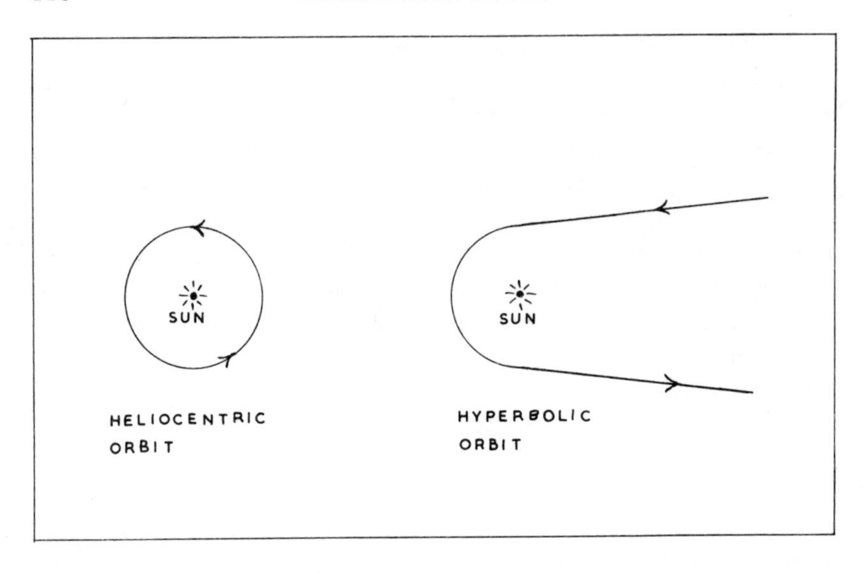


Figure 22

around planets or their satellites represent an infinitely better prospect. This still does not mean that detection of such objects would be an easy matter. Much would depend on their size. The material of which they were made would also be of importance, for the more of the sun's light they were able to reflect the easier they would be to detect. A final decision on the status of such an object would, however, depend in the final analysis on an on-the-spot investigation. A very long time could elapse before we are in a position technologically to achieve this. It might be felt that space vehicles of the Mariner type could, by close approach, be able to photograph or televise probable orbiting artifacts. In their present form, however, such craft could perform this function adequately only if the artifact's dimensions were fairly considerable. This might be the case were the artifact some form of space station. So far as small objects are concerned—and this seems much the greater possibility—our techniques in this field will have to be greatly refined and improved before we can usefully employ them in artifact hunting.

To sum up, then, we can say that our moon and the satellites of the giant planets are among the most likely spots for artifacts, especially the moon, since it is much closer to the center of the solar system. Of the planets themselves Mars and Mercury are probably the best bets followed by Earth and perhaps Venus. Orbiting artifacts are possible but great difficulties exist with regard to detection and identification.

Distinct ethical issues might arise were we to discover objects which clearly and unmistakably had been left behind by past expeditions from the stars.[5] These do not arise if the objects concerned are small defunct items—or even large defunct ones. Suppose, however, that in a secluded crater on the far side of the moon, or half buried within the sands of Mars, or orbiting peacefully around Ganymede we find an automated radio relay station which examination shows to be an integral part of some interstellar electronic network or an essential beacon for the ships of some galactic federation. Obviously we would want to examine it with infinite thoroughness. To do so would almost inevitably render it inoperative, probably forever. Have we a right to tamper in this way? Some might argue that the thing was trespassing in our system and since we are lords of that system we are free to blow it to smithereens if we are so minded. There is the other viewpoint. Do we terrestrials own or have dominion over the moon, Mars, or Ganymede? In the case of the moon the answer by now is probably yes. The position seems a lot less certain so far as the other two are concerned.

The question could go further however. Have we a right to tamper with things such as marker beacons? "Most assuredly," would be the reply of many, "for we do not want aliens, who could easily be hostile, coming this way!" But if we render inoperative one of their links, might this not be the very thing to send them probing suspiciously in our direction? The entire problem could be less concerned with ethics than with security. Admittedly, such arguments have almost science-fiction-like connotations at the present time. Perhaps the eventuality will never arise, but can we be absolutely sure of this?

By way of interest we should before closing this chapter point to an interesting hypothesis recently developed by a Scottish writer and amateur astronomer, D. A. Lunan, who claims that a probe injected into the solar system thousands of years ago by the people of a planet orbiting the star Epsilon Boötis picks up signals from radio stations on Earth and retransmits these back to Earth.[6] The result is a delay of several seconds between initial reception of the signals here and those allegedly retransmitted. Lunan has plotted the delay times graphically and the result is a surprisingly accurate representation of the constellation Boötes. Certainly instances of delayed radio signals have been occurring since 1928. Lunan's paper on the subject is extremely interesting and though inevitably it has aroused considerable controversy it is well worth reading and pondering over. The alien "transmitter," though an artifact of sorts, would represent not something left behind by an alien expedition but an automated object sent into our system in lieu.

Here, then, is another thought for us as we gaze into the depths of the night sky and watch the moon and planets silently and inexorably rolling on their predestined ways. Are there other things up there too? Could they be watching, listening, waiting?

# 10. BEFORE THE DAWN OF HISTORY

Earth had been born. Its infancy was over. But it was a wild, forbidding, and tempestuous place. Life too had been born but many rungs in the long ladder of evolution were yet to be climbed before the appearance of the great lord Homo sapiens. In the seas swam great and terrible fish. On the land, several rungs back on that same ladder, life had come ashore. Gills had been augmented by lungs and at length lungs alone sufficed. The creatures concerned were now marooned, had become wholly and permanently terrestrial. Swamps, grassland, forest, and jungle were henceforth their domain. Amid mountain chains and on island arcs in the oceans, great volcanoes hurled ash and rock explosively into the air.

Wild though it was, Earth was settling down. The aspect presented by continent and ocean was still far from that of the present day, yet a perceptive observer removed to these distant times could hardly have failed to notice that the continents, drifting gradually apart from an earlier contiguous land mass, were beginning to assume something of their eventual, familiar configuration.

Over a mountain range flanked on one side by thick, luxuriant rain forest and on the other by shimmering blue ocean the crippled starship spirals down. The sleek, sophisticated creation from the world of another and older sun has come from the rim of the universe, through a time-destroying dimension of terrible darkness. It has cleaved a path through the strangest of cosmic webs to be greeted in the end by the rays of a peaceful, yellow, and maturing star which a race yet unborn will one day call the sun. But in cleaving that path disaster swift and terrible has struck. Strange forces long enslaved, utilized, held in check, have burst free in one awful paroxysm of pent-up fury. Radiation and heat have seared the bodies of the strange creatures, so that now it is a dying ship manned by a dying crew. Dying eyes see impenetrable jungle beneath. Dying hands steer desperately for the one remaining source of salvation, the open sea. With skill born of long training and infinite experience, augmented now by terrible desperation, the failing, faltering ship is coaxed over the jagged mountains. It sweeps low through high passes, cols, and valleys, the clamor of its great motors echoing like a strange and terrible thunder. At last it is over the ocean. Lower it sinks, the noise from it now a muted whisper. Lower still and yet lower. Skimming the water now. Clouds of steam, spray, convulsive movements from both craft and water. It settles into its unnatural element. For a moment or two it floats, but its riven, strained, and tortured hull has reached and passed the limit of endurance. Water gushes torrentially into the great vessel, swirls madly and tempestuously around a maze of pipes, ducts, tubes, and wires. Resignedly its commander slams a roll of strange parchmentlike material into a gleaming metal tube which hermetically seals itself. Eventually, he hopes, the ocean bed which is soon now to be the grave of his ship, his hopes, his fellow beings, and himself, will one day, an infinity of time hence, be dry land. And then perhaps representatives of the race whose planet this becomes will find the remains. In this cylinder, if they are of sufficient intelligence, they will find the answer to many of their questions.

But it is not to be. The great vessel from the stars sinks quickly to the seabed and is gradually, over the years, covered by silt and mud. Centuries pass until eventually it is completely entombed. No longer do the waters of the ocean and the fish within that ocean play around it; no longer does the pale greenish-blue light of the sea filter through the clear transparent material of its few unriven ports. It is in its grave, the ocean floor and ocean far above it. Perhaps if fate or whatever else controls the destinies of these things had allowed it to fall in another part of the developing planet or into a different ocean the hope of its dying commander might have been realized. But this ocean floor was quite the wrong place, for in truth it was but a great conveyer belt supplied with insatiable devouring power by the strange, inexorable forces of crust and mantle. This ocean bed was moving and advancing remorselessly to meet it—a whole continent!

And so with the passing of millennia the entombed, encapsulated starship is carried toward the approaching land mass. Then, almost at the point of meeting, it is borne downward into a trench without bottom. Twin jaws of continent and ocean bed gradually grind it to terrible and total destruction. The mangled, utterly unrecognizable remains are carried inexorably downward into a realm of great and increasing heat. Rocks become plastic and melt, fusing into their midst all that remains of the once proud star vessel. The atoms of its substance, now at last free, join with those of their surroundings, eventually to rise up through the rocks of the continental land mass and be blasted skyward from the heaving throats of great volcanoes.

Alien relics and artifacts brought to our planet might easily have arrived long before man or even his earliest ancestors appeared on the terrestrial scene. The age of the Earth is currently believed to be of the order of 4,000 to 5,000 million years. Man has been around for only a very minute fraction of that immense period. Indeed if we care to represent Earth's age as 24 hours, man has appeared only within the last minute. During the almost countless million years of the Paleozoic, Mesozoic, Tertiary, and early Quaternary geological ages, a wandering ship or ships from a remote alien world might have spiraled downward through the fetid early atmosphere in a reconnaissance of a primeval, steaming Earth.

It has often been suggested, not altogether unreasonably, that if such visits were ever made, certain solid and enduring artifacts would surely have been left behind, intentionally or otherwise. Indeed, the question has been raised of a crippled starship still existing somewhere in a sort of fossilized condition deep within the rocks of our planet. Though the possibility of visits early in the geological past cannot be ruled out, the chances of remains proving sufficiently enduring seem very remote indeed. The geology of our world militates against it. Much depends on just how long ago a crippled alien vessel used Earth as a refuge—and in which part it descended. But assuredly if the event took place in Paleozoic times or earlier the chances of anything tangible remaining today are exceedingly slim.

To appreciate and understand this aspect of the question we must make a brief excursion into the realms of geology and geological history, paying particular attention to tectonic shift, or continental drift, to give it its more popular title.

In the opening paragraphs of this chapter we, in imagination, watched the dramatic descent of a great but crippled starship, its demise, and ultimate total obliteration. What precisely did happen to this hypothetical, unfortunate vessel?

When the crust of our planet first solidified and became reasonably stable the continental mass and the oceans presented a vastly different appearance from what they do at the present time. With the passage of thousands of millions of years the great land mass (which has been given the name Pangaea) broke up, the individual sections of this monstrous jigsaw going their own separate ways eventually to become the familiar continents of today. Now this incredible process is still at work and though throughout recorded history the continents appear to have remained stable and unchanged this is not really so. Our continents are still moving and will almost certainly continue to move. The rate of movement is, of course, very slow. For the record, the Americas on the one hand and Europe and Africa on the other are drifting apart at the rate of 4 centimeters per year, or about one mile every 40,000 years. This is hardly a prodigious rate in terms either of the human life-span or of human history. It is, nevertheless, very substantial on the geological time scale.

To sum up briefly, new crust is being created at mid-ocean ridges where basaltic lavas, for reasons still not wholly understood, are welling up from the Earth's upper mantle. As this magmatic material continues to rise previously ejected material is pushed outward and away on either side. To geologists and geophysicists this is known as a constructive margin—a region wherein new crust is continually being formed. A very typical example and probably the best known is the famous mid-Atlantic Ridge, which runs down through the center of both the north and south Atlantic oceans. For most of its very considerable length it is entirely submarine but it appears above the surface at several points, notably Iceland, an island renowned for its volcanic activity. In a lesser way it shows itself again as the Azores and as the island of Tristan da Cunha, both centers of vulcanism.

What happens now at the opposite extreme? The Earth is not becoming noticeably larger so it seems perfectly reasonable to assume that the formation of this new crust is being cancelled out by the destruction of old crust somewhere else. This, as it turns out, is precisely what *is* happening and for this reason such regions are known as destructive margins. At a destructive margin basaltic crust (formed aeons ago at a mid-ocean ridge) plus all the accumulated sediments and debris of the ocean floor comes into headlong collision with an advancing granitic continental mass. The latter overrides the marine crust, which thereupon dives deep down *under* the great granitic mass of the continent and back into the very mantle from which it emerged all these countless millions of years beforehand. In so doing some of the top layers are "scraped off" the marine crust to form new folded mountain ranges. The marine basaltic crust now plunging back into the mantle experiences tremendous heat and pressure which leads eventually to igneous and volcanic action on a fairly profuse and massive scale. The lava from such volcanoes is generally a blend of the basic basalt of the former ocean bottom and the acid granite of the continental land mass. Such lavas are termed "andesites" after the Andes Cordillera of South America, which represents a typical example of such a feature. Here the oncoming Nazca plate of the eastern Pacific, moving in an easterly direction, meets and plunges under the westward advancing South American continent. Just off the western seaboard of South America we find the typical deep ocean trench marking the region where ocean floor plunges downward toward the mantle. The highly folded Andes Mountains are noted for their very considerable and ofttimes violent volcanic activity and the names of these peaks roll off the tongue like a list of great warriors—Chimborazo, Canhuariazo, Quilotoa, Cimbal, Tungurahua, Aconcagua, Antisana, Cayambe, Sangay, Cotopaxi. None of these great volcanic peaks, it might be added, is less than 17,000 feet in height and in fact the soaring, snow-crested Chimborazo reaches nearly 21,000 feet.

By now the reader may well be asking what bearing all this has on the subject of alien visitors and visitations. Let us return to our imaginary and unfortunate starship. It fell into the sea and almost at once sank to the bottom. As century followed remorselessly on century it gradually became covered with silt. After many thousands of years it has become completely encapsulated so that in time the ocean bed lay not under it but *over* it. But all this time, slowly and inexorably, the "conveyer belt" of the ocean floor and the crust beneath, including the entombed starship, was continuing toward its eventual meeting with the juggernaut of the advancing continent. In the fullness of time that portion of the seabed plus the remains of the space vessel were drawn into the great yawning ocean trench and borne down under the enormous mass of the continent. Merciless jaws of rock ground the vessel from the stars into tortured, distorted fragments of metal. Pressure and tremendous heat fused the pathetic, unrecognizable remains. Molten metal mixed inextricably with molten rock. Some of this strange amalgam rose into the continental mass, solidifying within the mountains as intrusions of granitic, metalliferous rock. Some continued or was thrust farther until it reached the magma chambers of great volcanoes from whence it was eventually spewed as molten lava or hurled skyward as ash, fine dust, and hot gas. In any event, nothing could have been more completely destroyed, more reduced to its basic elements, than the luckless starship and the little that remained of its equally luckless occupants.

Now we are not trying to suggest that present-day volcanoes are all busy vomiting up the remains of alien star vessels. What this does illustrate is the fact that any alien vessels, or any alien artifacts for that matter, deposited on the Earth's surface in geological times (which represent the major portion of Earth's existence to date) would in most circumstances have been completely destroyed or utterly obliterated one way or another. If any descended on our world in these very remote prehistoric times and for any reason were unable to leave again, it is most improbable that their remains today, even if found, would be recognizable as such.

In early geological times (and in some areas in late) volcanic, seismic, and tectonic activity were much more rife and violent than they are today. It is difficult, for example, to conceive of a stranded starship or large alien artifact surviving an encounter with a massive flood of lava from an erupting volcano or the outpouring of magma from great fissures in the Earth's surface. For all we know, many times during the millions of years of the Paleozoic period strange vessels may have spiraled down through our skies. The wild primeval nature of our planet at that time plus its probable primary, non-oxygen atmosphere is hardly likely to have tempted their occupants to linger and even less to set up permanent or semi-permanent settlements save in an emergency. And as we have seen, the chances of remnants of these surviving to this day and age are, to say the least, remote. Even were whole ships to have been encapsulated in the rocks which formed around and above them they would almost certainly continue to be screened to all eternity from prying eyes and inquiring minds, for the rocks of the Paleozoic are very old and, except in instances of uplift or considerable erosion, are not readily exposed.

Such relics (or parts of them) just might come to light during deep coal mining operations. So far (and as far as is known) none has, but then decay and paucity in numbers could explain this. Most of the coal measures were laid down during the Carboniferous and Permian ages, both part of the Paleozoic period. Over a large portion of our globe at that time there was a monstrous "explosion" of plant life amid tropical, humid, and swampy surroundings—or more correctly, there were alternations of swamp and marine conditions. Vegetation grew luxuriantly, the remains rotting in the swamps. There came geological shifts and the swamps sank, the sea pouring in to cover them. Later the land rose again, new swamps formed, and the cycle was repeated. This shows all very clearly in the record of the rocks—layers of coal (fossil remains of plant and swamp life buried under anaerobic conditions) alternating with shale, sand, and limestone denoting the succession of swamp, delta, beach, and seabed.

It is unlikely, then, that we can anticipate anything positive pertaining to alien visitations during geological times, especially from the earliest of these periods. And even though such visitations did take place, the beings and their ships probably departed as they came, leaving no artifacts or other evidence of their transient passing.

There is another highly intriguing possibility that is raised from time to time: that unwittingly the aliens did indeed leave something behind—the germ of life which developed, multiplied, and grew on the surface of the hitherto sterile Earth. What can be said of such an idea? It is both interesting and fascinating, for, if it were by any chance valid, then all of us would in a sense be descendants of a race that exists, or once existed, at some unknown point among the stars of the Milky Way. Whether we would resemble them or whether the environment of Earth would have molded us in a different evolutionary form, we could never hope to tell. Indeed, were it true, the race from which we sprang might have become extinct millions of years ago.

There are few (and the writer is not numbered among these) who are prepared to give this idea much credence. Admittedly, the initiation of life is not an area in which science can as yet be very categorical. Just what spark, force, or ray at some point in time and matter changed an inorganic mass into an organic broth we do not know. A number of factors have been considered as possible initiators once nature had got its sums and building blocks right. There must have been many false starts. Radioactivity in the rocks, electrical discharges in the atmosphere, volcanic action—all are seen as possible initiators. It may even have been a blend of these. Currently, electrical discharges hold pride of place. Several interesting and revealing experiments have already been carried out which would indicate that amino acids can be synthesized from purely inorganic mixtures by such means. Amino acids are the building blocks of protein and in fact, as a study of the DNA helix shows, are absolutely fundamental to life. If life were formed in this way (or in some way approximating pretty closely to it) then it started as single-celled organisms—blobs of elementary, primeval slime (a sobering thought to reflect that these are probably our true progenitors and not godlike supermen from the stars). It seems reasonable to believe that such a process occurred (and is still occurring) in every "mother earth" in the universe where conditions are appropriate.

It is obviously much more rational and logical to accept such a beginning to life. Why, after all, if conditions were right should our planet have remained sterile, to be populated only as the result of a chance event by other beings from a world or worlds where sterility has not applied. Of course, life on Earth could already have been developing in its own indigenous way when the visitors arrived. The seed they unwittingly left could then have gradually become integrated into the developing and encompassing terrestrial biological structure. This concept seems just a *little* more reasonable.

Here again, there is no evidence, at least no obvious evidence, of an alien life-form, animal or vegetable, having found a niche in Earth's biological scheme of things. On the other hand, the evolutionary record of life on this planet from its genesis is far from a complete book. At best it is a number of incomplete chapters, a few scraps from other chapters and from the earliest barely a single word. Biologists and zoologists have traced back as well as they are able for long periods in the present and recent geological past. Their work can probably be considered reasonably complete and accurate. They have done a good job and have overcome many daunting handicaps. What, however, of the earlier geological periods: the Tertiary, Mesozoic, and Paleozoic? The easy answer is to state blandly that here the paleontologist takes up the task using the record of the rocks as told by the fossil remains within them. It is here that the greatest gaps take place. No shame to the paleontologist. What he has attained is remarkable in view of the difficulties besetting him. A visit to the fossil section of any good geological museum might give the impression that the record is very complete. Nothing could be farther from the truth. The gaps in our knowledge here are truly immense and though occasionally small chinks in the fabric are filled very large gaps remain and perhaps always will. Much of the fossil record has been destroyed by earth movements of varying types, by marine inundation, by erosion, and by igneous and volcanic action. And some, of course, is missing simply because fossilization did not occur in the first place. For the process of fossilization to occur there must be a certain rather unique set of circumstances—chemical, physical, and biological as well as geological. Such circumstances may be more the exception than the rule. As often as not there is no fossilization—just decay, decomposition, and ultimate disappearance. In the earliest of all geological times (characterized by the rocks of the pre-Cambrian) there is practically no fossil record of note—largely because the structure and form of the rather primitive life of the period rendered it incapable of fossilization even in the most favorable circumstances. Blobs of jelly, which is largely what composed the fauna of that very remote period, do not fossilize easily.

Even with a better fossil record we might not be very much farther forward. Life-forms of a basically different character often tend to converge. Convergence in a biological sense simply means that unrelated creatures tend to develop common characteristics because of their mutual environment. Take, for example, the common bat. It appears to fly—after a fashion. It even appears to have wings. In fact it is in many respects remarkably birdlike. Yet it is no bird, as a quick and cursory examination will reveal. The "wings" of the bat have little in common physiologically with those of a true bird, being in essence a form of leathery membrane attached to what are really the fingers of its anterior limbs. Genetically, too, the bat is a mammal and bears a reasonably close relationship to the rodents (mice, squirrels, etc.). Zoologically speaking, it is far removed from true bird. In the bat we have an original climber which liked to jump from branch to branch and then from tree to tree. Eventually it found that gliding (helped by an understanding and evolutionary nature) was the easiest way to achieve that method—hence the pseudo-wings.

It is probable, therefore, that a form of alien life deposited here aeons ago would also show convergence—to develop along lines dictated by Mother Earth despite a genetic "printed circuit" inherited from a far-off world.

It is most probably among plants that the idea of accidental alien implantation might have a little relevance. Here again there seems little possibility of ever differentiating between indigenous and alien strains. One could draw up a list of apparently unorthodox plants and wonder if in one or some of these we are seeing part of the flora of a far-off world. This is a tempting line of thought but one must be very careful. Take each plant on its merits. Examine its characteristics and, above all, its characteristics in the *light of its environment.* Let us, for instance, dwell for a moment or two on some of the odd—one might almost say fantastic—plants to be found amid dense selva jungle, e.g., the great rain forests of the Amazon. These are simply plants which have learned to survive amid great competition—plants that have been able successfully to adapt themselves to the peculiar and pressing conditions of their environment. To most of us they look alien, but they are almost certainly as terrestrial as ourselves. Whatever their odd appearance they are not descendants of species from another planet. The huge cacti of arid desert regions are also odd and grotesque by any standards. Indeed, few forms of plant life on this world of ours look more alien, yet we know quite well that these are merely plants which have become amazingly adapted to the rigorous conditions which surround them—so adapted in fact that an excess of moisture can prove positively harmful.

For the record, there are presently on this planet a total of no fewer than 328,610 *different* species of plant life. This includes all algae and bryophytes (mosses, etc.), although these amount to only 31,700. The main preponderance is of angiosperms (flowering plants), which represent 286,000 of the total. It is extremely difficult to construct a truly natural classification system for either plants or animals, for it must of necessity reach beyond observations of presently living organisms. Such a system must take into account what happens to be known of the evolutionary history of organisms, but because of the enormous gaps in the fossil record there is much uncertainty about lines of descent. Any system that mirrors evolutionary relationships accurately will be exceedingly complex for the simple reason that evolution itself is a very complex business. We must always bear in mind that organisms (plant and animal) may resemble each other *either* because of a common ancestry and hence evolution along similar lines *or* because they have come to resemble each other during the course of evolution from different ancestral stock. In circumstances such as these we can appreciate how impossible it would be to identify a point at which a non-indigenous organism intruded upon the scene.

This more or less brings us back to square one. All we can really say is that the idea of terrestrial life having had a non-indigenous origin is one that should, indeed must, be treated with considerable skepticism and caution despite its highly intriguing connotations. We must suppose there is just a chance of this having happened albeit an exceedingly remote one. The concept of an animal or vegetable form of life of alien origin accidentally finding a niche in the biological framework of our world is perhaps a little less remote though it is something we are unlikely ever to be able to prove either way. Should the rocks or the ocean floor ever reveal an artifact that is clearly and unmistakably of nonterrestrial origin and sufficiently old, then presumably the chances of some form of alien life having taken root here would be somewhat enhanced. It would not be proved. All we would know for sure was that once aliens set foot on our world. They might or might not have left the germs of life here—and even if they had, these germs might not have survived.

Despite all these difficulties and doubts there *is* a single fact that some are tempted to regard as odd. This is the advent of hominids, which from a strictly evolutionary point of view seems to have happened with peculiar suddenness. On the other hand, some very recent discoveries of sub-men cast doubts even on this. Was the appearance of hominids merely a delayed parallel line of mammalian development, a branching off from the hitherto mainstream, or was it due to something entirely different? This latter thought may from time to time nag and intrigue but in the light of present evidence (or, more correctly, the lack of it) it is simply not justifiable to proceed beyond this point.

# 11. IT'S IN THE BIBLE

At first sight the pages of the Bible might seem rather an odd place in which to meet references to possible alien visitations of our planet. A little reflection may temper this belief. The Old Testament of the Bible is not only a religious book (more correctly a series of books) but a record of much of the very early history of the Near East. To an overwhelming extent it bears the imprint of its own time. The translation to which many of us are accustomed was made about three centuries ago and inevitably much of the phraseology and idiom rings rather oddly in our ears. In places too the translation is less than perfect and this adds to the effect. To browse through the pages of a modern translation is something of a revelation and though no doubt the accuracy of the record is thereby enhanced, in some respects it appears to have lost a certain indefinable something. However, so far as our theme is concerned these are digressions.

The Bible, the sacred writings of both Judaism and of Christianity, is almost certainly the most influential collection of books in the world. The books of this anthology—for that in a sense is what it really is—were written over a period of many centuries, just how many has never been definitely established. The original languages were Hebrew, Aramaic, and Greek. The time span covered by the bulk of the Old Testament is in the region of one thousand years. Most archaeologists and historians agree that the Exodus occurred sometime after 1300 B.C. and this serves as a useful dating reference.

From a strictly historical point of view there is a certain lack of continuity. Nevertheless, the Old Testament is a magnificent historical record and contains much of considerable interest to historians and scholars as well as to theologians. We must always remember, of course, that it does not represent the history of the world up to that time—only the world of the eastern Mediterranean, parts of North Africa, and what are now the modern states of Israel, Lebanon, Syria, Iraq, Jordan, and Iran. The civilizations of the Far East and such other civilizations as then existed were to all intents and purposes unknown.

If aliens alighted on other parts of our planet, any record of the event, or events, will be that much harder to come by. But if any such event took place in the world of the Old Testament, then there is the possibility of mention although inevitably this will be couched in oblique terms. And if such visitations were as furtive (or sophisticated!) as modern UFOlogy would seem to suggest, then all manner of peculiar implications will be given to events which men of that time could only regard as manifestations from Heaven. It is therefore no use searching through the Old Testament for clear and unambiguous references for they will certainly not be found. One must read between the lines, ponder over the ornate language describing specific incidents, and having done so, weigh up the pros and cons. What was the precise nature of the event described? Can it be explained in a rational way—or is there something odd about it? Is an ancient chronicler trying to describe an event beyond his comprehension in the only language and by the only mental process he knows? No completely clear answer is ever likely to emerge. As often as not, mystery is heaped upon mystery. Nevertheless, a number of events in the Old Testament are highly intriguing and well worth some thought and study.

Probably the best place to start is with the Book of Ezekiel for here are described a series of events of a very significant kind. Indeed, in our context, this is probably the most significant part of the entire Old Testament.

First, a little about the Book of Ezekiel. There is some doubt about when it was written, although it was probably around 590 B.C. Ezekiel himself, probably a member of a family of priests, had been carried into exile in 597 B.C. His home was at Tel-abib near Chebar, which lay to the south of Babylon. About Ezekiel the man we do not know a great deal. He had been married but his wife was now dead, and from all accounts his family enjoyed a certain degree of importance. His age at the time he commenced his famous chronicle was probably about thirty but he was almost certainly nearer fifty when he made his final prophecies. There seems little doubt that by the standards of his time he had enjoyed a reasonably good education. The place and time of his death are not known although it may have taken place at a spot called Al-kiff about thirty miles south of Babylon, where the tomb of an unknown, which may be his, still exists.

And now to the book itself and the odd manifestations which Ezekiel so colorfully and eloquently describes. In our context a great deal could be made of the entire book and a little embellishment and imagination could make much out of the later chapters. This is a temptation to which the writer does not intend to succumb. Various interpretations could be put on these later chapters, some astonishingly favorable to our theme. It seems preferable, however, to restrict our speculation to what seems reasonable. We will therefore confine our attentions to the first chapter of Ezekiel, verses 1-28, and portions of the second and third chapters. There is much here which commends itself to our particular interest and even a cursory perusal of the passage should strike a significant chord. First, let us read chapter 1 in its entirety. The reader will find it rewarding to follow the passage slowly, thinking about the content and probable meaning of each verse.

**1:1** In the thirtieth year, in the fourth month, on the fifth day of the month, as I was among the exiles by the river Chebar, the heavens were opened, and I saw visions of God.

**1:2** On the fifth day of the month (it was the fifth year of the exile of king Jehoiachim),

**1:3** The word of the Lord came to Ezekiel the priest, the son of Buzi, in the land of the Chaldeans by the river Chebar; and the hand of the Lord was upon him there.

**1:4** As I looked, behold, a stormy wind came out of the north, and a great cloud, with brightness round about it, and fire flashing forth continually, and in the midst of the fire, as it were gleaming bronze,

**1:5** And from the midst of it came the likeness of four living creatures. And this was their appearance: They had the form of men,

**1:6** But each had four faces, and each of them had four wings.

**1:7** Their legs were straight, and the soles of their feet were round; and they sparkled like burnished bronze.

**1:8** Under their wings on their four sides they had human hands. And the four had their faces and their wings thus:

**1:9** Their wings touched one another; they went everyone straight forward, without turning as they went.

**1:10** As for the likeness of their faces, each had the face of a man in front; the four had the face of a lion on the right side, the four had the face of a bull on the left side, and the four had the face of an eagle at the back.

**1:11** And their faces and their wings were spread out above; each creature had two wings, each of which touched the wing of another, while two covered their bodies.

**1:12** And each went straight forward; wherever the spirit would make them go, they went, without turning as they went.

**1:13** In the midst of the living creatures there was something that looked like burning coals of fire, like torches moving to and fro among the living creatures; and the fire was bright, and out of the fire went forth lightning.

**1:14** And the living creatures darted to and fro, like a flash of lightning,

**1:15** Now as I looked at the living creatures, I saw a wheel upon the earth beside the living creatures, one for each four of them.

**1:16** As for the appearance of the wheels and their construction: their appearance was like the gleaming of a Tarsis stone; and the four had the same likeness, their construction was as though one wheel were within another.

**1:17** When they went, they went in any of their four directions without turning as they went.

**1:18** The four wheels had rims; and their rims were full of eyes round about.

**1:19** And when the living creatures went, the wheels went beside them; and when the living creatures rose from the earth, the wheels rose.

**1:20** Wherever the spirit would make them go, they went, for the spirit made them go; and the wheels rose along with them; for the spirit of the living creatures was in the wheels.

**1:21** When those went, these went; and when those stood,

these stood; and when those rose from the earth, the wheels rose along with them; for the spirit of the living creatures was in the wheels.

**1:22** Over the heads of the living creatures there was the likeness of a firmament, shining like rock crystal, spread out above their heads.

**1:23** And under the firmament, their wings were stretched out straight, one toward another; and each creature had two wings covering its body.

**1:24** And when they went, I heard the sound of their wings like the sound of many waters, like the thunder of the Almighty, a sound of tumult like the sound of a host; when they stood still, they let down their wings.

**1:25** And there came a sound from above the firmament over their heads; when they stood still, they let down their wings.

**1:26** And above the firmament over their heads there was the likeness of a throne, in appearance like sapphire; and seated above the likeness of a throne was a likeness as the appearance of a man upon it above.

**1:27** I saw as it were gleaming bronze, as the appearance of the fire round about enclosing him. Upward from what had the appearance of his loins, and downward from what had the appearance of his loins, I saw as it were the appearance of fire, and there was brightness round about him.

**1:28** Like the appearance of the bow that is in the cloud in the day of rain, so was the appearance of the brightness round about. Such was the appearance of the likeness of the glory of the Lord. And when I saw it, I fell upon my face, and I heard the voice of one that spoke.

What are we to make of all this? Here is a fascinating and intriguing account of something very odd, indeed. Not unnaturally Ezekiel attributes divine origins to all this. Ezekiel, we must remember, was a prophet in hot desert lands. His thoughts and words bear the unmistakable imprint of both the place and the time. An alien spacecraft (if this it was) was not something he could possibly understand. The possibility would not even suggest itself. In that age it was an unknown concept. To him it was simply an act of God.

In 1:4 Ezekiel says that a "stormy wind came out of the north," that there was a "great cloud, with brightness round about it." He speaks of "fire flashing forth continually." Now imagine some kind of spacecraft descending. There is a sudden shattering roar in the sky. Ezekiel looks up. To him the sky is opening. He sees lurid flashing flame gushing from amid a cloud of condensing gases and as the latter disperses the form of the descending machine begins to take shape. But Ezekiel would not think of a machine. Such a concept could have no relevance whatsoever. This was, must be, could only be a *divine* manifestation. He would therefore expect to see figures of one sort or another and in that welter of smoke, flame, and vapor he imagines he sees just that. He speaks not of one figure but four. He tells us they had four faces and that each of them had four wings, that their legs were straight and the soles of their feet round. Moreover, legs and feet sparkled like burnished bronze.

Suppose that some of us today were asked to describe the same event and that we were totally ignorant of all aspects of space technology. We would not find it easy and indeed the resulting description might be something of a classic! This very thought struck the writer when, in July 1971, as one of the observers of the Apollo 15 launch, he witnessed the dream of ages come true—men leaving for another world. Had he been transported not just 4,000 miles from Europe to Florida's sun-bathed shores but from a point in time several centuries before, how could he have described it? (For the record he is still far from sure!)

Let us continue then to analyze that valiant description by Ezekiel as he wrestles verbally (and no doubt mentally) with the seemingly inexplicable. An alien starship has approached our planet, though it is most unlikely that *this* is what Ezekiel saw. Huge sophisticated space vehicles are much more likely to go into a parking orbit several hundred miles above the Earth than risk the dangers of a descent. Again of course much depends on the technology of the aliens. There may be a parallel here with the huge ocean liners of the present time. If the port at which they are calling is too small or too awkward, those going ashore are put into tenders. The leviathans themselves are not going to risk it. The "tenders" for large interstellar space vehicles will be shuttle-craft of one form or another. It is therefore much more probable that what Ezekiel witnessed was the descent of a relatively small shuttle-craft emanating from a mother ship in a parking orbit several hundred miles above the surface of our world. As this shuttle enters our atmosphere it uses its jets for braking purposes. (These must have been very primitive aliens. Our science fiction friends would have opted for anti-gravity screens!) This seems a reasonable enough interpretation of Ezekiel 1:4.

Now Ezekiel begins to discern parts of the machine itself. Because he is totally uncomprehending he speaks of "faces" of "burnished bronze" and of "legs." What is he seeing? A descending ship using rockets or jets raises a tremendous cloud of vapor which may well obscure the approaching ground completely. In the circumstances now that the shuttle is well within the thicker layers of the lower atmosphere its occupants elect to use vertically mounted airscrews beneath it. Dust still rises in clouds but this is considerably less obscuring (what very primitive aliens—no automated system!). Alternatively the airscrews could have been employed as stabilizers. When Ezekiel speaks of "wings" he is probably referring to the spinning airscrews and the sunlight impinging upon them. He may even have been observing that rather peculiar but common effect whereby, despite rapid revolution, it appears as if a wheel or propeller is in fact revolving much less slowly in the reverse direction (1:9). (Remember the spokes of stagecoach wheels in Western movies?) The airscrews are mounted on vertical shafts so that they are spinning in the horizontal plane and apparently there are four of them (1:6). Apparently, too, they are mounted fairly close to one another—of necessity on a small craft. Because of this Ezekiel tells us that the wings "touched one another" (1:9, 1:11).

The vertical undercarriage legs have already come down—one at each corner of the craft, a wheel of sorts at the end of each (1:7, 1:15). The lightning mentioned by Ezekiel (1:13, 1:14) may have been brief spasmodic bursts of the rocket motor to assist the airscrews in slowing the descent. When he speaks of "burning coals of fire" this may be a reference to the jet orifices glowing red. By now the craft is close to the ground and he is seeing everything more clearly.

At last the machine has landed. "I saw a wheel upon the Earth," writes Ezekiel (1:15) "beside the living creatures, one for each four of them." In the prophet's mind, wings, legs, lightning, fumes, light, and shadow all add up to faces and living forms (1:10-15). Now details of the undercarriage begin to emerge (1:16-21), notably of the wheels: "their appearance was like the gleaming of a Tarsis stone" (1:16). Tarsis stone is another name for beryl, a bluish-green mineral. From this we must assume that the metal of the wheels was of a bluish-green hue but whether this represented its actual color or was due largely to reflections is hard to say.

The description accorded to these wheels by Ezekiel makes it possible they may have been metal *spheres* rather than wheels. To change their direction of travel wheels must reorient themselves. Spheres, on the other hand, merely revolve in the required direction. They may also have provided more than just a simple landing facility for which conventional shock absorbers would have been entirely adequate. The provision of wheels/spheres would imply that the craft was meant also to possess a certain mobility on land (1:17). Unlike the wheels of a conventional aircraft these were probably powered like the driving wheels of a car or truck. They have, therefore, treads to give them grip and these treads have bright metallic studs to augment that grip (1:16-18).

Ezekiel is very explicit about the wheels (1:10-21) and indeed this particular part of his fascinating chronicle approximates more closely to a present-day description than any other—"when the living creatures went, the wheels went beside them; and when the living creatures rose from the earth, the wheels rose." Not much dubiety about that: "Wherever the spirit would make them go, they went, for the spirit made them go . . . for the spirit of the living creatures was in the wheels." Thus in Biblical language we find the idea of living beings controlling wheels—in fact, controlling a machine.

Compared to the undercarriage the body of this craft is bulky, large, and perhaps hemispherical in shape; hence the appropriate description of 1:22: "Over the heads of the living creatures there was the likeness of the firmament, shining like rock crystal, spread out above their heads." Engines have stopped and the airscrews are now still. Ezekiel notes this fact (1:23): "Under the firmament their wings were *stretched out straight,* one toward another and each creature had *two wings* covering its body" (emphasis added). (Could this be a description of two-bladed airscrews?) Next he has more to say of the sounds—what it was like when the engines were running (1:24): "And when they went, I heard the sound of their wings like the sound of many waters, like the thunder of the Almighty, a sound of tumult like the sound of a host." This is followed by a reference to the stopping of the motors—"when they stood still, they let down their wings."

It is now that Ezekiel sees into the body of the machine and perceives a man. The reference this time is clearly to a man and not just to a creature (1:26): "And above the firmament over their heads there was the likeness of a throne, in appearance like sapphire; and seated above the likeness of a throne was a likeness as the appearance of a man upon it above." Perhaps hatches have opened or a sliding partition has rolled back. One of the occupants, perhaps its commander, is seen sitting upon a chair or dais. Not unnaturally, Ezekiel calls it a throne. By now, overcome by what he has seen, the prophet falls to the ground in a gesture of supplication. Whatever his emotions, his powers of observation do not seem to have failed him. In 1:27 he tells of the light surrounding this godlike creature: "I saw as it were gleaming bronze, as the appearance of fire round about enclosing him . . . I saw as it were the appearance of fire, and there was brightness round about him." No doubt in such a craft the lighting would be excellent—perhaps even by our present-day standards. To poor Ezekiel a mere bank of fluorescent tubes or a colored neon sign would have been miraculous. Presumably, the lighting was not monochromatic for in 1:28 he writes: "Like the appearance of the bow that is in the cloud on the day of rain, so was the appearance of the brightness round about. . . . And when I saw it I fell upon my face. . . ." In what other fashion could a man of Ezekiel's day and background describe colored lights and dials on a control panel?

From the account it would appear now as if the man on the "throne" addressed Ezekiel, for the final sentence of 1:28 runs: "And when I saw it, I fell upon my face, and I heard the voice of one that spoke."

The second chapter opens with this man telling Ezekiel to rise so that he may speak to him. There appear to be no linguistic difficulties. According to Ezekiel, the monologue that followed dwelt chiefly on the sins and shortcomings of the children of Israel.

At this point the prophet was probably so overcome by a mixture of awe and sheer physical shock that his senses would interpret any words he heard, even in an unknown alien tongue, as the only possible type of message such a godlike creature descending from the heavens in a cloud of smoke and fire would utter. In 2:9-10, the closing verses of the chapter, the man is clearly showing something to Ezekiel: "And when I looked behold a hand was stretched out to me and lo, a written scroll was in it. And he spread it before me; and it was written within and without. And there was written therein lamentations and mourning and woe."

The third chapter, verses 12-15, strongly suggests that Ezekiel was then taken by the craft to another place.

> Then the spirit lifted me up, and as the glory of the Lord arose from its place, I heard behind me the sound of a great earthquake.
>
> The spirit lifted me up and took me away and I went in bitterness in the heat of my spirit, the hand of the Lord being strong upon me.
>
> And I came to the exiles at Tel-Abib, who dwelt by the river Chebar. And I sat there overwhelmed among them seven days.

Again, despite the Biblical language and metaphor, this sounds precisely the way, the only way, in which a holy man, a prophet, could possibly describe such an experience. Just what the effect must have been on the "exiles at Tel-Abib" if he arrived in their midst in this fashion is best left to the imagination. It is hardly surprising that he sat there "overwhelmed among them seven days." The exiles must have been equally numbed!

All kinds of questions arise regarding this strange incident.

Was Ezekiel in some kind of trance or did he dream it all? This is a reasonable and logical question and probably the first to arise. This might be the case except for the nature of the description. Even clothed in its quaint Biblical language it is clearly a description of an aerial craft. How could a man in those far-off days possibly imagine technical features of the sort of which he speaks? What type of trance or dream could possibly achieve this? Even if Ezekiel were only a storyteller drawing on a vivid imagination he could hardly have conjured up things like these. Ezekiel *saw* something! Of that there can be very little doubt. And that something was remarkable. Such a machine could not possibly have existed upon Earth at that time. It could only have come from regions outside our world. Were these regions extraterrestrial—or extra-solar? Could there have been a civilization on Mars which has since perished? If not, we must look to the stars, which certainly adds a new urgency and realism to the theme of this book.

Are we, on the other hand, misconstruing it all? Are we reading into the Book of Ezekiel things that are not really there? Again, a thoroughly reasonable question, but once more we come up against those uncanny descriptions. Will there ever be an answer to the great enigma of Ezekiel?

In the Book of Exodus, chapter 25, we find a further perplexing matter. It is here that God gives Moses precise instructions for constructing the Ark of the Covenant. The specifications are fully detailed and very exacting. In fact, another two chapters are required before they are complete. On several occasions Moses is warned to make no mistake in the design. Clearly this is vital.

The real mystery and it is a considerable one, comes later. We find this in the Second Book of Samuel, chapter 6, verses 1-11. David had decreed that the Ark be moved from Gibeah to Jerusalem. It was carefully and reverently transferred to a new cart driven by one Uzzah. Approaching journey's end some passing oxen brushed against the cart almost overturning it. Seeing disaster about to overtake this most holy and venerated of relics, Uzzah rushed to steady it. He dropped dead instantly! It was as if he had touched some highly electrified object.

Perhaps he had! If one reads carefully through Exodus 25 and correlates what he reads with the basics of static electricity it becomes rather obvious that the Ark was an electrical capacitor capable of taking and holding a charge of several hundred volts! The metals brass and gold abound within the thing, the actual capacitor plates being of gold—one set having a high negative charge, the other an equally high positive one.

Now it must be obvious that the ancient Israelites had not the knowledge to construct lethal electrical capacitors. That discovery lay two to three thousand years in the future. Did they manage it by accident? Most improbable! Read Exodus 25 again. The instructions are clear and precise; there is nothing haphazard about them. From whence, then, came that knowledge? "From God," many may reply. We would not like to dispute that. We would only ask, "What agent did God use to impart this knowledge?"

One of the most perplexing disasters recorded in the Old Testament is probably the sudden and utter destruction of the twin cities of Sodom and Gomorrah. This is told rather briefly in Genesis, chapter 19. The Biblical account is a fairly well known one. In Genesis 19:12-14, we are told that "angels" told Lot to take himself and his family out of the cities with great haste for the latter were doomed and would very soon be completely destroyed. Apparently the family were not convinced of this impending disaster for the following morning they were still there. The "angels" appeared again, urging haste, but still Lot and his family lingered. At length, we are told, the angels more or less forcefully removed Lot, his wife, and his two daughters from the danger zone. Having conducted them some distance in the direction of safety they leave them with this admonition (19:17): "Escape for thy life, look not behind thee, neither stay in the plain; escape to the mountain lest thou be consumed." Whatever the identity of the "angels" they clearly knew what was coming. Despite objections by Lot they hurried him along. One cannot avoid the impression that they knew exactly when the cities were to be destroyed.

And then it happens (19:24-25): "And the Lord rained upon Sodom and Gomorrah brimstone and fire out of heaven; And he overthrew these cities and *all the plain,* and all the inhabitants of the cities and that which grew upon the ground" (emphasis added). In reading this one gets a mental picture not just of Sodom and Gomorrah but of two much later-day cities somewhat farther to the east! Obviously Sodom and Gomorrah were completely obliterated along with all the surrounding region. The "brimstone and fire," of course, we must simply regard as Biblical language. These were the only words the ancients knew to explain something so cataclysmic.

There are further points of interest. Had the twin cities been destroyed by fire (which was the utmost such a people could achieve by way of destruction) there would have been no need for

Lot and his family to have retreated as far as the mountains. Even the destruction of the cities by modern high explosives would not have necessitated this. Not only did the "angels" know of the coming blast and of its precise time. They knew also of its cataclysmic nature. They knew something else besides. That something is of considerable significance today. Consider just four words in Genesis 19:17: "Look not behind thee." Since 1945 all of us have been well aware that unless the eyes are well protected one does not look directly at the blossoming of an atomic "sun." There is an equal awareness of the dangers from radiation, which is yet another extremely good reason for putting a fair bit of territory between oneself and an impending nuclear blast. Even the B-29 *Enola Gay* did not hang around after she laid her "atomic egg" on Hiroshima.

There would then seem to be some evidence to support the idea that Sodom and Gomorrah were destroyed by a nuclear explosion. What relevance this has to their alleged wickedness we do not know. It may be that these now legendary cities were no better and no worse than all others at that time—only rather less fortunate.

Suppose we examine the record of the episode from a point hours before the great blast. In Genesis 19:1 we find Lot sitting at the gate of Sodom. Here he is approached by two strangers. They are clearly very distinctive and different in appearance. Lot regards them as angels and bows low before them. He invites them to stay the night in his house but the strangers refuse. Lot pleads with them to do so and eventually the strangers relent (19:3): "And he pressed upon them greatly, and they turned in unto him and entered into his house; and he made them a feast . . . and they did eat." Apparently Lot's strange visitors had not gone unnoticed by the citizens of Sodom, for in verses 4-5 we find a mob milling around Lot's house demanding to "know" the strangers. Lot refuses. The mob continues to press him, but Lot remains adamant. He will not accede to their demands. The position becomes quickly more acute. Now in a desperate attempt to protect his strange visitors he is driven to offer his two daughters to the mob (19:8): "Behold now, I have two daughters which have not known man. Let me, I pray you, bring them out unto you and do ye to them as is good in your eyes; only unto these men do nothing. . . ." The Bible does not record the reaction of the two young women to this proposition! The offer is ignored by the crowd, the mood of which has now turned very ugly (19:9): "... now we will deal worse with thee than with them. And they pressed sore upon Lot and came near to break the door." Now, and none too soon, the strangers decide to take a hand in the proceedings (19:10-11): "But the men put forth their hand and pulled Lot into the house to them and shut to the door. And they smote the men that were at the door of the house with blindness, both small and great...." What was the strange power of these men that they could blind an entire crowd? What manner of weapons were they employing?

It is at this stage we learn of their mission—and of its strange terrible urgency (19:12-13): "And the men said unto Lot. Hast thou here any besides? son in law, and thy sons and thy daughters, and whatsoever thou had in this city, bring them out of this place. For we shall destroy this place...." It is now that Lot's family refuse to accept the situation. Presumably they are unwilling to believe such a fantastic prediction. This is still the position the following morning when the urgency of the situation compels the strangers to take the forthright action we have already mentioned. Lot and his family are thereby saved—all, that is, except for his wife, who looked back "and became a pillar of salt" (19:26). In fact, she "looked back from behind him"—a rather significant detail. Just what precisely happened to this luckless woman is not certain but it would appear that she lingered behind the others to look back and was still within range of the radiant heat wave when the holocaust took place. Her extinction would appear to have been fairly instant. The effect of the blast was certainly total (19:27-28): "And Abraham got up early in the morning to the place where he stood before the Lord; And he looked toward Sodom and Gomorrah, and toward all the land of the plain, and beheld, and, lo, the smoke of the country went up as the smoke of a furnace." This could almost be a description of Hiroshima that fateful August morning in 1945.

At that time in history only a natural cataclysm such as a huge meteorite fall, a great earthquake, or a violent volcanic eruption could have caused so great and so sudden a degree of destruction. All these can fairly certainly be ruled out since it is clear that prior knowledge of the event was possessed by the strangers. We must assume that the strangers or their fellows were responsible for the catastrophe—and surely only a nuclear explosion could have wrought such havoc. If it were such an explosion, then it must have been engineered by beings from beyond this planet. The general impression given by the strangers is that, though different in garb and demeanor they must have borne a close resemblance to terrestrial man. How they were able to converse in Lot's tongue if they were not of this world is impossible to say. Perhaps they never really did converse other than by signs and gestures.

If the supposition that these strangers were aliens is correct, then why the blast? Had their craft landed and developed some reactor defect? Perhaps the reactor had begun to run "wild" and they had no means of bringing it under control. Perhaps they had a load of fissionable material in a highly dangerous state and it had become imperative to dispose of this. In view of the experiences suffered by the strangers while in Lot's house they may well have come to the conclusion that the occupants of Sodom and Gomorrah deserved all that was coming to them!

Chapters 1 and 2 of the Second Book of Kings is also worthy of close study. Verses 10-14 of the first chapter are especially interesting. Verse 10 reads as follows: "And Elijah answered and said to the captain of fifty. If I be a man of God, then let fire come down from heaven and consume thee and thy fifty. *And there came down fire from heaven* and consumed him and his fifty" (emphasis added). By the time the end of verse 14 is reached two more batches of fifty and their captains have been effectively incinerated. Is this just an example of picturesque language, of extravagant speech, dubious translation—or *did* fire come down from heaven? If so, what was the nature of that fire? Lightning perhaps? A very convenient electrical storm if it was. And if it was not lightning—?

All this took place on Mount Carmel around 896 B.C. and it would seem that the heat developed was considerable. Oddly enough, there are reports of blue crystal fragments being dug up on the slopes of Mount Carmel in recent years. These are thought to be about three thousand years old. The interesting fact is that chemically identical material was found at Alamogordo, New Mexico, just after the first atomic device was detonated. Prior to the test no such material had been found there. This, of course, should be interpreted only as a *possible* link. Natural glass (obsidian) emanating from certain volcanic regions, though normally black, does occasionally appear in a colored form due to the presence of metallic impurities. Glass was also made, albeit crudely, in the remote times of which we are speaking. Normally glass requires a temperature in the range 1300-1600°C for its formation from silicate sand. This is considerably less than that produced near the heart of a nuclear explosion. The significant feature of the material found on Mount Carmel (and in the New Mexico desert) is that it does not fuse (melt) at temperatures as low as these!

Consider also II Kings 2:11: "And it came to pass, as they still went on and talked, that behold there appeared a *chariot of fire* and horses of fire, and parted them both asunder; and Elijah went up by a whirlwind into Heaven" (emphasis added). How would, indeed how *could,* an ancient chronicler describe an aerial or spacecraft any other way? We are not seeking to suggest that a spacecraft *was* definitely involved. Nevertheless, the thought is unavoidable.

Another intriguing possibility is the destruction of the city of Jericho in 1451 B.C. According to the Bible, Jericho's walls and buildings collapsed as a result of a massed trumpet blast by the investing army. For obvious reasons this is seen as improbable. If these massive walls *were* toppled by means of acoustically generated vibrations we must certainly wonder where the necessary equipment to achieve this originated. Assuredly it was from nowhere on this planet.

A German archaeological expedition in the late twenties claimed to have found evidence that Jericho's odd downfall was brought about by extreme heat. This has been substantiated by a British expedition that examined the remains in 1930 and reported "traces of intense fire, including reddened masses of brick, cracked stones, charred timbers and ashes. . . ." From the British account it would appear that the heat generated was considerable and probably in excess of that to be expected from a normal fire. On the other hand, a burning city can generate tremendous temperatures if a number of small fires link up to create what is known as a "fire storm." Several German and Japanese cities during World War II experienced this dreadful phenomenon after bombing. Nevertheless, these modern cities were large conurbations set alight by sophisticated incendiary devices. It is less easy to envisage a small ancient city suffering a conflagration of this nature by the means available then. Can we, in the circumstances, think of the explosion of a nuclear device or the use of some kind of heat ray, perhaps of a laser nature? Just who did destroy Jericho—and how? In Joshua, 6:21-24 we find confirmation of the use of fire but no more: "And they utterly destroyed all that was in the city. . . . And they burnt the city with fire, and all that was therein. . . ."

If by some strange chance nuclear war did occur in these ancient times, certainly the peoples of Earth did not supply the means. If these means came from beyond, then today we have cause for disquiet.

The Bible also makes predictions. In Peter II:3-10 we find the following: ". . . the heavens shall pass away with a great noise, and the elements shall melt with fervent heat, the earth also and the works that are therein shall be burned up." Is this a warning of another invasion from the skies? Of course by now man is crazy enough to achieve all this by his own hand—a sobering reflection!

So far as the Bible is concerned and especially the Old Testament, one could go on and on. It is a very mysterious collection of chronicles. There is, it must be admitted, a temptation to put an extraterrestrial connotation on many of the quaint and seemingly inexplicable happenings. This is a trap into which many writers with the best of intentions have already fallen. One must beware. So easily can wish become father to the thought. The instances quoted here are the most significant and so far as the Old Testament goes probably the best documented. It seems inadvisable to proceed further in this particular direction. Nevertheless, if the reader has the time and the inclination, a thorough perusal from Genesis to Malachi could yield much food for thought.

# 12. TERRESTRIAL—OR ALIEN?

It is tempting to look around at some of the monuments, statues, carvings, and other relics, large and small, bequeathed to us by many centuries of recorded history—and sometimes of unrecorded history—and attribute to them extraterrestrial origins. There do, certainly, exist a few examples of an unusual nature for which it is difficult to deduce valid, rational explanations. This does not mean that no such explanations exist. Perhaps so far we just haven't thought of them. But on the other hand . . .

During the course of the last few years a number of writers have devoted entire works to this admittedly intriguing theme. Most are extremely interesting and at times highly ingenious. On the whole, however, they are not tremendously convincing. Not infrequently, speculation is extrapolated to hyper-speculation. One does not doubt the sincerity of these authors and even less their enthusiasm. As often as not, it is this very enthusiasm that engenders these extrapolations. Enthusiasm given a free rein can lead to many odd conclusions, especially with a subject of this nature. Unfortunately, this can do a great deal more harm than good. In this way the perpetrators largely throw into disrepute the very cause so dear to them.

The opposing attitude is the one most likely to be held and expressed by professional astronomers, who, because of the strict disciplines of their training, react with considerable, and on the whole, commendable caution.

This other point of view was expressed by Dr. A. E. Roy, Professor of Astronomy at the University of Glasgow, who wrote, "The super-beings, being incredibly careless with their super weapons and super technology, left artifacts here and there, we are told, scattered about our planet, and it is only the hidebound, reactionary attitudes of orthodox scientists, archaeologists and anthropologists that prevent acceptance of this new evidence of our origin by celestial midwife." [1]

In Chapter 9 we saw how artifacts would be much more likely to endure in some parts of the solar system than in others, and in Chapter 10 we came to realize how our restless Earth could so easily and effectively have obliterated all traces of alien visitation or presence. Therefore, even if there is no clear, unambiguous evidence of such occurrences on our planet, this should not be construed to mean that they did not take place—only that evidence of them *may* not have proved sufficiently enduring. Of course, this line of reasoning must not be seen as some kind of proof that aliens *have* been here. The entire argument represents an extremely slippery slope and it is at all times necessary to proceed with the utmost caution.

However, artifacts left on the Earth by aliens during the last two or three thousand years *could* have survived under favorable circumstances. The older the artifact the more favorable would these circumstances require to have been. It must also be apparent that an artifact of a more durable material such as hard stone or nonreactive metal is likely to survive longer than one fashioned from plastic or other synthetic material. In short, the circumstances are all-important.

Compared to the geological epochs, which are measured in hundreds of millions of years, two or three thousand years are as nought, and during such a "short" period artifacts are unlikely to have been destroyed by processes of tectonic shift (continental drift) unless they were deposited too close to some spot where this was already active.

All this is reasonable but it is still necessary to counsel caution. In each and every instance considered we *must* first of all investigate the more reasonable possibilities. Even a somewhat improbable orthodox explanation is more likely to approximate to the facts than a "way-out" extraterrestrial one. We must also take into account the verdict of professional archaeologists and anthropologists. Where the layman might so easily be baffled experts could come up with perfectly legitimate and reasonable possibilities. Only when all these avenues have been explored are we entitled to turn our thoughts toward these certain "other possibilities." Even then, we must proceed with the utmost caution because time, weather, and general decay could so easily have destroyed evidence of a vital character. The lack of this might send us off merrily down a totally false trail.

Our entire approach has to be one of compromise. The subject of alien artifacts is intriguing, fascinating, and, highly emotional to some. In recent years it has come to interest many people in all walks of life and of all ages. Since representatives of our race have walked upon the moon, mankind has become much more acutely aware of the immensity of the universe, of our small place in it, as well as of the scope and power of space technology. No longer is space travel something from the pages of a bedtime novel or an hour's pleasant excitement at the neighborhood movie theater. Much of science fiction is in the throes of becoming science fact. And since men of Earth can now walk upon the surface of another world (albeit a nearby one) might not men and women from more distant worlds walk upon the surface of Earth?

We must at all times be very careful that subconsciously we do not fabricate evidence with respect to alien artifacts. This in no sense is meant to suggest that all proponents of the idea are shameless charlatans. To do so would be grossly unfair. Nevertheless, auto-suggestion and self-delusion can be, and often are, powerful forces rendered even more potent by virtue of the fact that they are not always recognized for what they are. Many examples of this could be quoted but perhaps one selected at random will suffice. In many parts of the world can be found wall paintings, frescoes, and carvings which show among other things men wearing armor. Now it is common knowledge that throughout recorded history armor has come in many different forms, shapes, and guises. Inevitably, a few examples bear more than a passing semblance to modern space suits. How easy, then, to assume and proclaim that these *must be* ancient crude drawings and caricatures of astronauts who descended upon Earth in the remote past. While this may just conceivably be possible, it is quite illogical to state, suggest, or insinuate that they *are.* Some ancient armor has a vague space suit sort of look. But, of course, some space suits are a bit reminiscent of certain ancient armor! It works both ways. It might also be added that some of the figures in these ancient frescoes and drawings resemble modern deep-sea divers, yet it is hard to recall any instance in which the suggestion has been made that some ancient people on Earth made a point of exploring the seabed!

We must, therefore, beware of this sort of thing unless it can be backed up by other reasonable evidence. If it cannot, then a claim of the above sort will almost certainly result in the scientific establishment ripping it to shreds and gleefully dancing on the pieces. As we said earlier, there must be compromise. Let us seek the true mysteries, the few items that border on the inexplicable, and leave alone those things which, despite a certain aura of mystery, are much more likely to be terrestrial. This is only an application of something we mentioned before. In science undue speculation is to be avoided, but so also is a rigid, closed, inflexible mind. The correct pathway here may be far from straight, but assuredly it is *very* narrow!

The history of science reveals quite a few examples of the closed mind attitude. Charles Darwin discovered this to his cost, as did those geologists who claimed that the surface of the Earth was not all of marine origin. In 1790 the German physicist Ernst Chladni was the target for virulent abuse because he dared to suggest that meteorites were of cosmic origin; and about a century and a half ago it was firmly held that the ocean depths were totally unsuited to any form of life due to lack of oxygen. Nevertheless, we must not allow the errors of the past to push science from fact into fantasy. Science, however, has its frontiers and these have continually widened over the years. But they would hardly have done so if a number of dedicated men and women had chosen to regard them as forever unchangeable.

It has long seemed to the writer that evidence for extraterrestrial visits is more likely to take the form of small artifacts than large structures or complex equipment—at least so far as the surface of our planet is concerned. Much, no doubt, depends on how long our alien visitors (if any) stayed on our world. Despite the heroic endeavors of those writers who would have us believe aliens started our civilization it is surely more probable that visits were brief and desultory. This would not make them any less real or any less interesting. If aliens had spent long years on our world a few millennia ago, it stands to reason that evidence of their presence would be fairly definite. Despite this, the remains of ancient civilizations, notwithstanding many contemporary claims to the contrary, look convincingly terrestrial. A brief visit does not of course mean one of just a few hours. "Brief" in this context could legitimately be interpreted as several months or even a year or two.

Let us start with something that is not in the strict sense of the word an artifact at all—a footprint.[2] The reader may well say, "So what?" Well, this is a fossil footprint and it was found in rocks of the Tertiary period. Still nothing very remarkable, were it not for the fact that this was not an animal footprint! It strongly resembled a human one. At this time in Earth's history even our apelike ancestors had not appeared. An error of observation perhaps? Unlikely! With the exception of the very oldest rocks (the pre-Cambrian) geologists can date rocks with reasonable accuracy. But why then just *one* fossil "hominid" footprint? Surely there should have been others? This does not follow. Despite the host of fossils found in our rocks they represent only a very minute fraction of the creatures that once lived. For a fossil to be formed several essential conditions must be met—e.g., the place of death must be favorable, the form of the creature's (or plant's) structure must be capable of fossilization, the remains must be buried before decomposition gets under way. For the most part these conditions are only rarely met. Moreover, even after fossils have been formed, they can be destroyed by shifts in the Earth's crust, by metamorphism, and by vulcanism. For these reasons, among others, the fossil record of the rocks is very incomplete. Another factor applies in our case. A footprint is a kind of "negative" fossil and is more likely to be obliterated before fossilization can take place. When a creature or plant dies something solid remains, but a footprint is merely an impression, which can disappear on a single tide. Moreover, an alien's footprint would, we imagine, be rare, comparatively speaking. We can thus make some case for this fossil footprint's being alien. For the record, this particular find was made in the State of Nevada in the 1880s. An instance such as this seems more likely to support the case for past alien visitation than an involved harangue which seeks to establish, against all evidence to the contrary, that some ancient temple in the jungles of Southeast Asia is due to godlike creatures from the stars!

And so to another tiny artifact of considerable interest. This one, like the Nevada fossil footprint, is not strictly an artifact, either, though it may once have been. This is the imprint of a simple, homely object—a small screw just over 5 centimeters long—in the middle of a piece of rock millions of years old.[3] The screw itself had long since oxidized or corroded away but its "fossil" remained clearly in evidence. There seems no physical way in which a screw could possibly have found its way into the rock—and if this is so, then the screw must have been encapsulated long before a terrestrial civilization capable of making (or needing) screws arrived on the scene. From whence could it have come but from "above"? An obvious query would be whether this *is* in fact the imprint of a screw. Could it be a natural cavity in the rock which by some strange freak of chance resembles the "cast" of a screw? This appears very improbable since the impression is symmetrical and virtually perfect in every detail. Could it have been made by the shell of some marine creature living at that time? The writer has lying in front of him at the moment of writing the fossil shell of a gastropod. The dimensions are about right and the shell has a helical form, but the resemblance to a screw is very superficial and since fossil gastropod shells are common it seems virtually impossible that the "cast" in the rock could have been mistaken for such an object. Unless by some near miraculous means a screw passed down and into Tertiary rock, then it predates terrestrial civilization by several million years. This object, or more correctly, imprint, was also found in Nevada, a few miles from Treasure City, in the year 1869.

If perhaps readers find it odd that the Nevada fossil footprint was human in all respects we might now quote an instance of footprints which assuredly were not. Neither were they those of any known terrestrial creature. They were found near the source of the Tennessee River in solid rock. This time there was more than just a single print. Moreover, they showed evidence of *six* toes and were over 30 centimeters in width. The result of a check by the writer on his own "pedal extremities" showed these to be no more than 10 centimeters at their broadest and he has no reason to believe that his feet are in any way exceptional! Near to where these prints were found others were also discovered. These were clearly hoof marks but of considerable dimensions, 25 centimeters broad on average. What type of man—or ape—on Earth has ever had feet with six toes and a breadth of 30 centimeters? There seems no really logical answer to this.

There are reports (though so far the writer has not been able to secure reliable documentary confirmation of these) of a sarcophagus being unearthed near Crittenden in Arizona in 1891 which contained the remains of a "human" being about nine feet high, with *six toes* on each foot! Similar remains are reported to have been found in the Caucasus Mountains, though the reports from there do not speak of six toes. Still, a man with only five toes per foot *is* rather exceptional if he is nine feet tall. This does not, however, automatically elevate him to alien status.

There are also reasonably well-substantiated reports of a metal object found inside a chunk of quartz in California.[4] The object, whatever its original use or function, had obviously been machined and well finished. This was still apparent despite an age estimate of around twelve thousand years. It has been likened to a bucket handle, which in the circumstances seems a little misleading. If its origins are terrestrial, then presumably buckets of a sort could have been in use at that time, though hardly with handles of finely machined metal. If, on the other hand, it is an alien artifact, it seems insulting, even irreverent to imagine a sophisticated group of beings capable of transit between stars using something so mundane as a bucket. We must presumably assume that in this instance the "bucket handle" was not a bucket handle! At a place called Kingoodie in Scotland a very similar object was found, only this time it was encased in stone. The age was again put at twelve thousand years.[5]

Over the years quite a number of objects have been found amid stone and coal but the pedigree of most is highly suspect. In some instances the evidence of the hoaxer's hand is all too plain. Signs of improvisation are not very hard to find, especially for the expert.

There is on record the very interesting report of a bison's skull (prehistoric) discovered in Siberia. A neat round hole in the forehead bears a marked semblance to that likely to be caused by the entry of a high-velocity rifle bullet. We hardly need to ponder whether firearms existed then, so if the wound *was* inflicted by such a missile a very considerable mystery attaches to this matter. There would appear to be three alternative and strictly terrestrial possibilities here. One is that it is due to a meteorite. This seems so highly improbable that it is better dismissed. The second, that it was caused by some boring shellfish,[6] seems a little more likely, but if this is so, the bison must already have been dead when the creature fastened upon it. The third of these alternatives, that it was caused by some small dense solid object driven against the skull by gale-force winds, is feasible. Small objects in hurricanes have been known to drive deeply into solid objects, so clearly this possibility cannot be dismissed, especially if the skull were already old and the bone very brittle.

Since spears or arrows were the only weapons at that time it is not unreasonable to ask whether this hole might have been caused by such a weapon. This, however, would not have caused a smooth round hole. It must also be stated that the edges of the hole were calcified. It looks therefore as if the beast had somehow survived the wound.

Despite all we have said about this particular find there should be no undue haste to attribute it to aliens. All we have done is to contemplate a number of likely orthodox reasons, none of which seems all that convincing. There may still be a good and simple reason for the injury caused to this beast. Had aliens been abroad with such weapons one might expect other such examples to turn up but until now (so far as the writer is aware) none have. If the hole were *proved* to be due to an explosively fired missile we would almost certainly have to start thinking in terms of aliens. There would be one other essential—the hole would require to be of the same age order as the skull. We must never eliminate or forget the possibility of a hoaxer having found the skull, making a neat hole, then replanting the object. This is one of the inherent weaknesses of small artifacts. Faking a large alien shrine or launch pad would be considerably less easy—and presumably less tempting.

It is now about time to look at larger possible artifacts and we might as well start with something on the grand scale. This feature has exacted considerable interest: the gigantic linear patterns on the floor of the Peruvian desert near Nazca which are thought to be at least two thousand years old.[7] These are, in fact, vast geometrical shapes marked out on the desert. The lines are dead straight to the limits of measurement. Pottery fragments found beside them suggest the lines were carved out before A.D. 100. What was the purpose of this tremendous enterprise? First beliefs were that these lines might actually be sightlines for a huge astronomical almanac of the sun, moon, and stars. Gerald Hawkins of the Smithsonian Astrophysical Observatory has carried out a search for every conceivable celestial alignment without finding anything significant.[8] When these patterns are viewed from the air in their entirety there is a very uncanny resemblance to the system of runways, intersections, and perimeter tracks of a modern airport. We can certainly rule out natural origins for these amazing markings. But why should a race living almost two thousand years ago have gone to all this effort? Despite Professor Hawkins's failure to correlate the patterns with astronomical alignments we cannot abandon completely the theory that this was the purpose.

The symmetry of the patterns is such that one feels they must originally have been drawn out on a small scale and then reproduced on the desert. To achieve this, an exceedingly high expertise in surveying would have been necessary. It is hard to believe that this could have been possessed by a people living two millennia ago.

From ground level the system of markings is much less impressive. They still excite attention but not nearly to the same extent. They are found to be broad furrows which reveal the pale yellow subsoil lying beneath the brown sand of the desert. An immediate question is how they could possibly have survived the passing of some twenty centuries. The answer to this probably lies in the almost complete absence of rain in the region plus a peculiar kind of protective "skin" (probably an oxidized coating) that protects them from the wind. The Nazca plain is reasonably accessible since the Pan-American Highway runs across it, intersecting many of the straight markings.

Erich von Däniken[9] has suggested that the markings are not unrelated to the comings and goings of aerial craft. There is not much, however, to back up this assumption, though he cites a "trident" near Pisco 820 feet long with arms 12 feet six inches wide, pointing inland with two islands out at sea on its center line. This, he maintains, might have been a marker for aerial navigation as it points toward the Nazca system of lines. It might indeed, but on the other hand this could be merely a fortuitous circumstance. It could not possibly be accepted as proof.

Although the system has this amazing resemblance to the layout of a modern airport, we must beware of drawing the obvious conclusion. Seen at close range these broad, straight tracks could not possibly have accepted aircraft. Of course the possibility arises that if they *were* meant for aircraft such craft might not have been the heavy multi-wheeled giants now in vogue. A suggestion has been made that they were intended primarily for air-cushion landing craft.[10] But why, then, we are entitled to ask, should we have long straight strips? Surely craft of this type can set down almost anywhere. Indeed that is their overwhelming advantage.

A line drawn from the "trident" at Pisco to Nazca and then extended approaches the famous high-level inland lake of Titicaca in the Andes. Von Däniken points out that this constitutes virtually a straight line.[11] In fact, the three sites are closer to the arc of a great circle from which the orbital plane of a circling starship could be obtained.[12] Such an orbit would be inclined at approximately 35 degrees to the equator and a high-altitude site in the region of Lake Titicaca could be well placed for return launches to orbit. Again, this is highly interesting but unfortunately such speculations are impossible to prove. Once more it may be just an example of coincidence. This is a difficulty inherent in our subject. Intriguing possibilities arise and we speculate. But always we arrive at this irritating dead end.

The markings on the plain of Nazca are very odd and practically inexplicable in terrestrial terms. They point to very high standards in surveying which could hardly have been possessed by an ancient Earth race. Neither is their origin natural. Brains have conceived them, hands have made them, but from whence came these brains and hands? And certainly the system would appear to have "aerial" connotations. We cannot safely go farther than that.

We mentioned already that the Nazca patterns might have been sightlines for a vast astronomical calendar though evidence for this so far is distinguishable by its absence. Similar reasons have been advanced to explain other ancient monuments. There are, for instance, a number of stone monuments and alignments in Scotland and in Brittany which appear to be the remains of observatories for recording the moon's motion 3,500 years ago.[13] Indeed, recent archaeological research has shown that Megalithic man went to incredible lengths to solve the problem of the moon's motion and also that he had an astonishing knowledge of both astronomy and geometry.

A retired professor of engineering at the University of Oxford,

Alexander Thom, has spent a considerable number of years making accurate surveys of about four hundred Megalithic monuments throughout Britain.[14] The conclusion he has reached is that Neolithic man in the second millennium B.C. probably possessed a knowledge of lunar astronomy far exceeding that of his descendants over the ensuing three thousand years. It is, for example, possible to predict eclipse phenomena by making observations at Stonehenge, that famous circle of great stones on Salisbury Plain, Wiltshire, England. Professor Thom has also drawn attention to the fact that many of the less impressive Megalithic remains are of considerable astronomical importance. At several sites Neolithic man followed the moon with great precision and had developed a capability of measuring to within a few seconds of arc.

Professor Thom has been criticized by a number of archaeologists on the grounds that he is attempting to read too much from the stones, since surely any group of stones will show a pair of apparent astronomical alignments purely by chance. Professor Thom counters this by virtue of the precision of his measurements.

One highly interesting result of his work is evidence which would seem to indicate that the orbit of the moon has not altered during the past four thousand years. Some of these stones render possible calculations of the relative tilt of the orbits of Earth and moon with respect to the sun. Values thus derived are in close accordance with those obtained by contemporary methods. There would therefore appear to be little doubt that these ancient stone circles were constructed as astronomical observatories.

Apart from Stonehenge probably the most impressive of these monuments is that at Carnac in Brittany. Here the great stones are arranged not in a circle but in twelve great rows three miles long. Some of the stones are now missing—hardly surprising in view of the immense age of the thing. Professor Thom and his colleagues were able, however, to establish without doubt the sites of the missing members. It is already apparent that the object of this complex setup was nothing less than the solution of an astronomical problem by means of geometry.

At the center of the Carnac complex lies a massive block of stone now broken into four parts. Originally it must have been 67 feet long and weighed around 340 tons. As to how this stone in that day and age could have been moved to its present location one can only quote the words of Professor Thom: "a mute reminder of the skill, energy and determination of the engineers who erected it three thousand years ago."[15] This colossal block of stone, known as Le Grand Menhir, apparently served as a foresight or locus for lunar observations made by astronomers standing close to the surrounding megaliths. To position it with the necessary degree of precision the moon would probably have had to be observed for hundreds of years. Once again we cannot do better than quote the words of Professor Thom.

> At each maximum or minimum, parties would be out at all possible places trying to see the Moon rise or set behind high trial poles. At night these poles would have needed torches at the tops because marks would not be visible until actually silhouetted on the Moon's disc. Meantime some earlier observatory must have been in use so that erectors could be kept informed about the kind of maximum which was being observed. There would then ensue nine years of waiting till the next standstill when the other four sites were being sought.[16]

In the light of all this many questions arise. For example:

1. How did Neolithic man become so knowledgeable?
2. From whence did he obtain the engineering to move such colossal blocks of stone?
3. Why did he go to all this trouble?
4. Was it only Neolithic man that was involved?

Dr. Samuel Milton sums it up rather well:

> There still remains unanswered the question as to why they went to all this trouble. Certainly the desire to predict eclipses featured prominently in the minds of the builders, and at sites in West Scotland they were probably useful for indicating dangerous tidal conditions. But the sheer enormity of the projects at Stonehenge, Callanish and Carnac hints at *something deeper.* [Emphasis added][17]

It is difficult to equate Neolithic man with intellectual inquiry and achievement of the kind we have just been describing. It is equally difficult to do so with respect to the engineering necessary to move a great rock weighing over 300 tons. Yet it seems reasonably certain these things were done—and in different parts of Europe, perhaps even of the world. Was all this the work of Neolithic man or was some superior race involved? In this case the latter seems the more probably alternative, yet one hesitates to be categorical on such an issue. These megalithic remains pose a mighty question. As they stand stark, silent, and deserted at night against a pattern of the eternal stars dare we assume that between stone and star there is a strange, ancient affinity?

Other structural work attributed to Neolithic man has been investigated by Professor R. J. Atkinson of the University of Cardiff.[18] During recent excavations of Silbury Hill in southern England, it became apparent that the mound's constructors had an excellent knowledge of soil mechanics, for they had incorporated several series of concentric circular walls into the hill to prevent collapse of the earthwork.

It seems also that mining technology was highly developed in Europe as early as 4,300 B.C., flint mines dating from that period having been uncovered.[19] These early miners were apparently in little doubt as to where the best flint was located, for there is evidence of their burrowing through no fewer than ten inferior seams to reach the best.

Professor Thom has also investigated a stone circle in the Orkney Islands off the north coast of Scotland.[20] This is the Ring of Brogor and is the most perfectly constructed stone circle known. It has a diameter of 250 meters and consisted originally of 60 stones spaced at 6-degree intervals around the circumference. To lay this out without modern surveying aids must have constituted a most formidable task. Professor Thom believes this must also have been a huge lunar observatory, probably the most northerly in Europe. It is believed to have functioned around 1600 B.C., although there is no radiocarbon evidence to substantiate the claim.

Neolithic remains of the type we have been discussing provide considerable food for thought. This was the New Stone Age, the age which preceded the use of metals. It is distinguished from the older or Paleolithic Age by the more advanced workmanship, particularly the grinding and polishing of the cutting edges, as well as by the variety of form and use displayed by implements wrought out of flint and other hard rock. It was the age of lake dwellings, inhabited caves, and burial mounds, the age that saw the invention of basketry, weaving, pottery, and the domestication of animals and plants. Considerable progress has always been assoicated with this age—e.g., corn grinding and the development of stone and timber architecture—but it is difficult to equate such elementary development with the intellectual and engineering skills inherent in the remains we have been studying. Without doubt there is an element of mystery here!

A number of interesting relics in North America would also seem to indicate that the inhabitants of that continent had at one time also a surprising familiarity with the stars. A prime example is located in Wyoming and another in Saskatchewan, Canada. The former, on a remote peak in the Bighorn Mountains, is traced out in stone blocks and resembles a large wheel having 28 "spokes" and a diameter of 80 feet. The area was once the hunting grounds of tribes whose names are now part of the legend of the West: Cheyenne, Shoshone, Arapaho. According to archaeologists, the "wheel" had no connection whatsoever with these tribes and predates them by a considerable period. The alignments of six rock cairns just beyond the perimeter of the relic point to the solstice sunset and the rising positions of the three bright stars Aldebaran, Rigel, and Sirius. It would seem that the early inhabitants of these plains, about whom we have very little knowledge, had a surprising interest in the heavens. The similar pile in Saskatchewan spans 200 feet atop Moose Mountain. Here, too, the placing of cairns results in alignments with the summer solstice sunrise and the rising points of Aldebaran, Rigel, and Sirius. Archaeologists believe this was constructed more than a thousand years ago and that it represents a relic of an Indian heritage that has long been forgotten.

There is no suggestion that these peculiar objects were constructed by or under the direction of past visitors to our world. The connection, if any, lies in the acquired knowledge of the sky by an early and fairly primitive terrestrial people. Depositions taken from the region's occupants during this century and the closing years of the last recall little practical use of the heavenly bodies. Do these enigmatic stone circles represent the practical calendar of a people who seem to have been instructed in something more than the rudiments of observational astronomy? Again we can only wonder how and from where that knowledge was acquired.

Much can be deduced from ancient wall carvings so far as normal terrestrial events are concerned. It is also not very difficult to find items which might appear to have extraterrestrial connotations. Such speculation can be risky—many might even say ill-advised. There *are* suits of armor that look like pressure suits; there *are* birdlike figures bearing a strong resemblance to modern jet aircraft; there *are* odd-looking objects that might be firearms or lasers. Chances are, however, that these items are suits of armor, birds, and early weapons, respectively—no more and no less! We must at all times remember the myths of ancient civilizations. Much was made of religion and of "gods." Portrayal of these gods (or their envoys) would most likely be in the form of near-human creatures in the sky bearing wings. How easy to regard these as astronauts of long ago.

There is, however, one item which seems worth mentioning since its credentials are probably better than most. This constitutes part of an ancient Egyptian wall carving in the Temple of Hathor, near Dendera in Egypt. It depicts an intriguing scene: figures bearing giant vessels apparently of glass. Clearly visible inside these are long, thin, sinuous objects. The immediate impression is of huge electric bulbs complete with interior filaments. Moreover, the "filaments" are linked to what looks like a switch box or power source. Indeed, the lines connecting the "filaments" to the "box" are almost identical to certain heavy-duty cables used today. In addition, the "bulbs" are supported by objects which are an exact replica of modern high-tension insulators. Despite the great age of the carving it has apparently suffered little, thanks no doubt to the very arid Egyptian climate. It should be added that there are numerous references in ancient Egyptian records to supposed "gods" predating the first dynasty. These records refer to an early superior civilization possessing allegedly miraculous powers. Illustrations and photographs of this particular carving can be seen in numerous works and the similarity to modern electrical apparatus borders on the uncanny.

It is also of interest to examine the contents of ancient records.

Here one comes upon much that is highly enigmatic. How, for instance, could ancient people, long before the invention of the telescope, be aware that Mars was the possessor of two moons and Jupiter of four large ones? Yet records mention these facts. In some ancient Babylonian chronicles even the phases of Venus are described. If these were guesses they were highly accurate ones.

A book written in India about 3,500 years ago contains some very startling material. This is the renowned Mahabharata, a vast compilation of over 100,000 couplets. The central theme of the book is the struggle for supremacy between two rival factions and it is based on historical events that took place not later than the tenth century B.C. It also deals with the creation of the universe, legends, history, and the deities of ancient India. The work was first translated into English in 1884. There is much of interest. Consider, for example, the following passage:

> A single projectile charged with *all the power of the Universe.* An incandescent column of smoke and flame, as bright as ten thousand suns, rose in all its splendour. It was an unknown weapon, an iron thunderbolt, a gigantic messenger of death which reduced to ashes the entire race of the Vrishnis and the Andhakas. The corpses were so burned as to be unrecognisable. Their hair and nails fell out, pottery broke without any apparent cause, and the birds turned white. After a few hours all food stuffs were infected. To escape from this fire, the soldiers threw themselves in streams to wash themselves and all their equipment. [Emphasis added]

One can only inquire how a description of atomic warfare could have been written around the tenth century B.C. Even the *form* of the nuclear weapon is mentioned:

> A shaft fatal as the rod of death. It measured three cubits and six feet. Endowed with the force of a thousand-eyed Indra's thunder, it was destructive of all living creatures.

By this reckoning the weapon was over ten feet long.

There is even the graphic description of two bomb-laden aerial craft or nuclear-tipped missiles colliding:

> The two weapons met each other in mid-air. Then earth with all her mountains and seas and trees began to tremble, and all living creatures were heated with the energy of the weapons and greatly affected. The cities blazed and the ten points of the horizon became filled with smoke.

We can be quite sure that civilizations in the Indian subcontinent in that remote age did not possess a nuclear arsenal—nor even one of conventional chemical projectiles. Is this, then, merely a flight of fancy, an early example of science fiction, or did events of this traumatic nature occur? If they did, the weapons employed could not possibly have been of terrestrial origin.

Would not past nuclear explosions on Earth have left visible scars, vestiges of which might still remain? This is a difficult question to answer. Who today without prior knowledge could guess that Hiroshima and Nagasaki were destroyed in this way only three decades or so ago? Nature is a good and quick healer. Man is also fairly effective in clearing up the mess he has made. It is reasonable, then, to assume that such events taking place three thousand years ago have left no traces easily recognizable today. However, there could be another side to this. During archaeological digging in southern Iraq in 1947 a form of mine shaft was created. Proceeding downward this successively passed the levels of Babylonian, Chaldean, and Sumerian culture, of primitive agricultural communities of around 7000 B.C., and finally of cave dwellers of roughly 14,000 B.C. It was then that the archaeologists came on a layer of fused glass which resembled very closely a material found at Alamogordo, New Mexico, just after detonation of the first atomic weapon in July 1945. This can hardly be regarded as conclusive, but it is interesting.

The temptation to scan the history books in an effort to secure material that might have alien connotations is strong. The recent rash of books concerning artifacts, slanted legends, strange "gods," and so on has embraced just about everything from Atlantis to the Great Pyramid, from the Lost Continent of Mu to the Biblical Flood. Much is interesting—much should not be taken too seriously! It is one thing to ponder a possibility. It is quite another to state emphatically that a particular object here on Earth *is* the product of alien hands.

The overall impression given by these books is that this little planet of a very average yellow star has, over many centuries, been a source of interest to many galactic races, that these alien people have interfered with our history to no small extent, that no continent on our planet has escaped their attentions. Despite this, the evidence for their being here is rarely unequivocal. When we step back and have a good look at it the picture is not impressive and certainly not convincing.

The belief of the writer can be summed up in the following points:

1. Mankind is by no means alone in the universe.
2. Races millennia in front of ours will have perfected satisfactory techniques of interstellar travel.
3. Representatives of such races probably have on a few occasions come our way.
4. Such visits may have taken place before life appeared on our planet, during prehistory, and at odd intervals during historical times.
5. Oblique references to such visits and their effects may be found in ancient books such as the Old Testament of the Bible.
6. Aliens *may* have bestowed certain "know-how" on ancient peoples (e.g., Neolithic man).
7. Initial visits may generate return visits from time to time.
8. Return visits may be increasing because of the accelerating pace of terrestrial technology since the latter half of the nineteenth century.
9. Artifacts are more likely to be found on planetary surfaces where atmosphere and geology are less destructive, e.g., our moon, the larger moons of the giant planets, and perhaps Mars.
10. Alien artifacts with a continuing purpose—e.g., listening devices, relay stations, telemetry stations—may exist within the solar system in specific orbits—perhaps amid the asteroid belt or on the moon—for monitoring the affairs of Earth.
11. A high proportion of UFO lore is rubbish but a small significant fraction is not and should not be ignored.
12. Aliens might one day elect to visit us openly and in strength.

For the people of Earth this could be a dazzling prospect—or a very terrible one!

The writer believes these twelve points to represent a reasonable compromise between the views of those who would have us believe that mankind is either alone or unreachable and of those who would give an alien slant to just about every major event in history. Truth is generally found somewhere between two extremes.

# 13. REFLECTIONS AND RECOLLECTIONS

Before embarking upon the present-day scene a few personal experiences may prove of interest. The last few decades have certainly provided more than a little in this respect—everything in fact from transient UFO sightings to lurid accounts of encounters with alien beings.

For the record I have to state categorically that despite many years of intense interest in the heavens and of assiduous sky-watching I have encountered no little green men; no accommodating Martians or Venusians have offered me a free "flying saucer" trip to either of their respective home planets. And even if they had (and I had managed to survive the shock) I would have serious doubts about relating the fact in these pages since, quite understandably, few would believe me.

In this respect, then, I have not a great deal to offer—merely three minor incidents which may add up to very little. Nevertheless, I find them as intriguing and perplexing today as I did at the time each occurred. I must emphasize that I do not necessarily attribute "alien" origins to these events. I state only the facts as they are known to me. Readers will find no attempt at embellishment.

The first of these events took place on the evening of March 1, 1938. My interest in astronomy and matters celestial had been kindled only a couple of months previously. I was at the time a fifteen-year-old schoolboy who had not even as yet aspired to the luxury of a small telescope. Indeed, my entire equipment amounted to a set of star maps, a chart of the moon, and a pocket flashlight. I could hardly be considered a serious rival to Mount Wilson or Mount Palomar. Being without optical aid at that time was no real handicap since I was doing what every novice to the fascinating world of astronomy ought to—getting to know the constellations and how to find my way about the star-strewn depths of the night sky. This is a rewarding and enjoyable pastime and one highly commended to all lovers of the great outdoors.

By now I was fairly familiar with the star patterns visible at that time of the year. Already I had come to regard these remote, twinkling points of light as familiar and faithful friends. Looking to the north I could locate the stars of the Great Bear—known better perhaps to those in North America as the Big Dipper, a title which somehow conjures up visions of plains, pine forests and Indian encampments of long ago. There also was Polaris, the Pole Star, the Little Bear, and Cassiopeia. And sometimes too in that direction could be seen the shimmering shafts and streamers of the Northern Lights. In the south, striding across the winter sky in all its glory, was the magnificent Orion and around it the stars of Taurus and Pleiades, Auriga and Gemini. Low in the south, Sirius, the Dog Star, flashed resplendently. Familiarity with the summer sky, so far as I was concerned, still lay some months ahead. Then, during the long twilights and warm nights of that season, I would get to know the stars of Cygnus, Hercules, and Aquila and see the lovely Vega "burning" in its strange, far-off fire.

On that frosty, star-powdered night of March 1, 1938, I had been locating for the first time the stars of Corona Borealis (the Northern Crown) and Boötes (the Herdsman). The appearance of these constellations over the eastern horizon is a sign that spring is at hand in the northern hemisphere. That night, however, the feeling was still of winter and in fact a week later the landscape was once again clothed in a thick mantle of snow. Prior to retreating indoors for the night I had a last quick look around the starry vault above, reluctant to leave it. In southeastern Scotland it was a perfect night, a touch of frost, no moon, and the stars stood out in the most brilliant way against the sky of sheer black velvet. Indeed, one felt he had but to reach out to grasp the stars.

It was then, just as I looked to the north, that I saw it! Now in those days there were no such things as artificial satellites, and in that region and at that time night-flying aircraft were very much a rarity. So in that peaceful vault of the night sky nothing should have been moving—yet clearly and unmistakably something was!

North of Corona Borealis and Boötes and running between the constellations of the Great and Little Bears is a long straggly line of stars known as Draco, the Dragon. It is a very undistinguished constellation at the best of times and none of the stars in it are particularly bright. By now, however, I was completely familiar with each naked-eye member of that constellation. I was consequently more than a little surprised (which is putting it mildly) to note that in some peculiar manner it had recently acquired an additional member. For a minute or so I imagined this to be a nova and excitedly contemplated the measure of astronomical fame which must descend on a schoolboy observer discovering a new star! Such thoughts and hopes were quickly doused, however, for the star, "thing," or whatever it was, was moving among the stars—not quickly but with an easily perceptible movement. By now I knew that meteors did not move in this slow, stately, and measured way. I was equally aware that comets showed no movement—at least not over a period of a few minutes.

I continued to watch the object to be certain I was not imagining things. Clearly I was not. I saw the distance between it and a real star shrink, then it passed that star and began to close the gap between itself and another star. There was no possible doubt about it. As I continued my vigil the mysterious object moved slowly and deliberately in a northerly direction toward the stars of the Little Bear (Figure 23). Once or twice it seemed to wink but generally the light was steady. Now in 1938 all aircraft without exception were piston-driven and compared to today's machines flew at no great altitude. This meant that invariably one heard the sound of their motors. But from this object there came no sound and in the still, frosty air of a small country town far from large centers of population the sound of aircraft engines would most certainly have been heard. Then it entered completely into the Little Bear, blinked hesitatingly some six or seven times, and disappeared. Although I hung about expectantly (and hopefully) for about twenty minutes there was no repeat performance. Silently it had come—and just as silently it had gone.

At the time I had no opinions on the subject. It was to me (and still is) an inexplicable phenomenon. "Flying saucers," UFOs, and other weird forms of spaceborne crockery were then quite unknown so that any such ideas never entered my head. At that time in

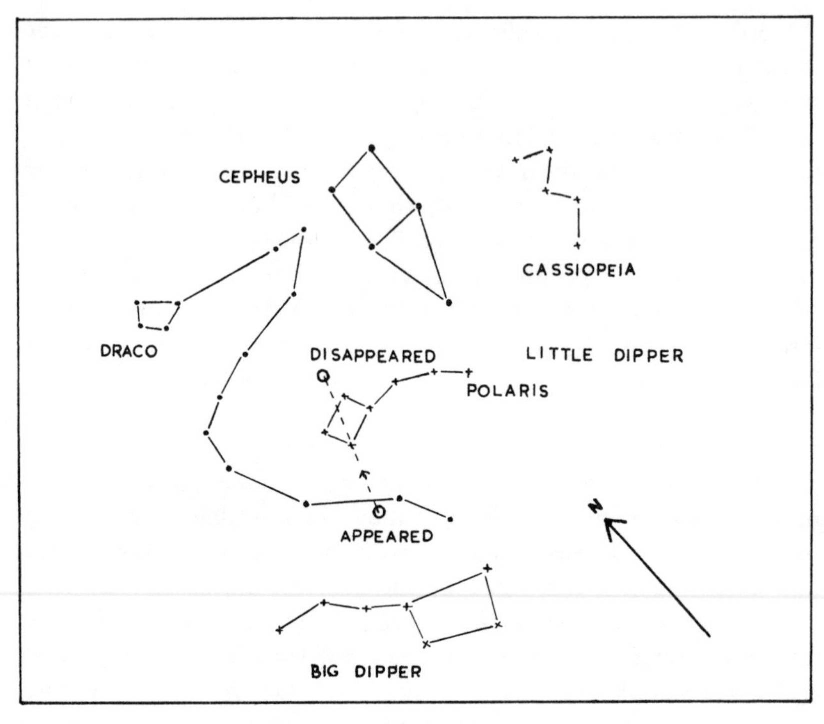


Figure 23

*Path of object seen from Kelso, Scotland, 21:00 G.M.T., March 1, 1938.*

Britain there were no domestic civil air services worth speaking of, especially in this part of the country, and though the beginning of the second world war was then less than two years distant, little was ever seen of military aircraft. And certainly they did not pass through the night sky without a sound blinking in this odd fashion.

As the years have passed and more and more has come to be heard of unidentified flying objects this inexplicable, intriguing experience has remained very clear in my mind. Perhaps there exists a completely rational explanation, but if one does I must admit that it has not so far occurred to me. If in fact any readers of these pages have ever experienced a similar phenomenon, especially about that time, I would be very interested and grateful to hear from them. I realize only too well, of course, that rather a long time has elapsed since 1938.

For the record I cannot say that I experienced any vague feelings of menace. This is a feature which nowadays is often mentioned.

Perhaps I might if UFO tales had then abounded—but these still lay a decade in the future. At any rate I did not feel that every shadow, every bush, every hedge harbored an alien monster that had descended silently from the skies even as I watched. That still belonged to the worlds of Flash Gordon and Buck Rogers, whose odd adventures could be followed from week to week with relish at the local cinema.

So far as I am concerned events now jump seven years—to the summer of 1945. A lot had happened since that night in 1938. World War II had come and by now had almost gone. I was with the Royal Air Force in India and on this particular day, as radio crewman on a Transport Command C-47, was returning to base at Agra after a routine haul to Nagpur. The flight was a normal, more or less everyday affair, and I doubt if any of that four-man crew were thinking about anything in particular unless it was a cool shower and a pleasant drink in the mess after landing. Our intention had been an early morning takeoff when the air was still pleasantly cool and free of "bumps." However, the renowned Law of Maximum Perversity had decided to operate in the shape of a highly reluctant port motor. Result—a three-hour delay in departure. By now the sun stood high in the kind of intense, brassy sky that only India can produce.

We were flying at an altitude of roughly 7,000 feet. The air by now was very bumpy and the ground, the broad, barren, and featureless plains of central India, lay sprawling and baked beneath us. Here was a dried-up riverbed, there a small village, and occasionally the gleaming white of a temple, shrine, or other religious edifice showed up clearly. We had seen it all before, too often in fact—and we were sick of it. The romance of the so-called Mystic East soon palls! Suddenly, apparently from nowhere, a seemingly solid oval object appeared beneath us. At first the contours of the thing were quite clearly defined but after about a couple of minutes they began to look a little hazy. Its color was also varying, going from heliotrope to pink, then back again. It seemed to be visibly pulsating. The most disquieting feature so far as we were concerned was the fact that the object was apparently matching its speed to our own. Regrettably, we had no cameras aboard the C-47 so a permanent record of the event was out of the question. Naturally we reported what we were seeing by radio but the news produced only ribald comment, which was more or less what we expected. The thing was with us for a full five minutes, then just disappeared as quickly as it had come. I would like to be able to report that the radio channels were full of weird alien gibberings and strange cadences but they were not. Apart from an unusually high level of static all was normal.

Later the meteorological people told us that what we had seen was probably a "mock sun" produced by the existence of a thin layer of ice crystals at an altitude lower than that of our C-47—a phenomenon which, though comparatively rare, is not unknown. Ice crystals at that height during the heat of an Indian day! It seemed odd. Still, I expect the Met people were right. They are supposed to know about these things. Nobody else took us very seriously and for that reason we let the matter drop. UFOs were then unknown, although the era was just about to begin. At times I still wonder about this incident.

My remaining personal experience of what might just conceivably have been some kind of UFO (and here again I do earnestly stress the word "might") took place during the evening of September 3, 1957. Now 1957 may not seem so very long ago and indeed it is not. Nevertheless, at the present rate of technological progress twenty years counts for quite a bit. At that point in time, space travel was still regarded by most people as something of a pipedream—the preserve of science fiction writers and addicts and not really respectable in the eyes of established science. Indeed a recently appointed British Astronomer Royal had only some eight months before described it as "utter bilge"! So far not a single satellite swung in orbit around the Earth, with the exception, of course, of a rather large one thoughtfully placed there aeons before by the Almighty! Another month was to pass before the launch of the Russian Sputnik I was to catapult our civilization, whether it liked it or not (and it probably did not), into the Space Age. Thus anything seen in the sky at that time could not have been an artificial satellite, launch vehicle, or spacecraft—unless someone had already achieved a launch in a very clandestine way.

The time was around 9 P.M. (British Summer Time). The evening was of that pleasant kind so typical of early autumn. The sky was completely cloudless (unusual in western Scotland) with the sun sinking gradually toward the western horizon. The room I happened to be in faces the northwest and from it one has an excellent view of the rugged mountainous terrain of the lovely island of Arran, which lies in the Clyde estuary some twenty or so miles off the coast of west-central Scotland.

At that particular moment my thoughts were far removed from UFOs, flying saucers, or any other form of spaceborne crockery. Indeed, just then the spate of postwar sightings had apparently diminished. Or maybe for the time being the news media had come to the conclusion that enough was enough.

My attention was drawn suddenly to an intensely bright, golden, starlike object which was sailing quickly across the sky on a steady, straight northwesterly course. At first I imagined it to be an aircraft. From my home the large international airport of Prestwick is only about fifteen miles distant. As a consequence most of the aircraft we see are fairly low-flying—either just beginning to climb or preparing to land. As with most areas, of course, the region is also overflown by high-flying aircraft both civil and military. Nevertheless, there was a certain indefinable something which seemed to render this one strangely different. For a start, it emitted no sound. On such a still and perfect evening there certainly should have been. Another unusual feature was the fact that there was no vapor trail. Neither of these points is particularly significant, though in view of the apparent altitude of this "flying object" I would have expected a vapor trail, especially in the climatic conditions then prevailing. Only a short time before a thoroughly conventional aircraft had traversed this same patch of sky and its very obvious vapor trail of condensed gases had endured for several minutes before dissipating. By a stroke of good fortune I had within easy reach a pair of fairly powerful binoculars. When I focused these on the object I saw quite clearly that it was no aircraft—at least not of the conventional kind. The thing gave the impression of being disc-shaped. No markings or external detail were visible. There was a dead calm that night—in fact we were slap in the middle of a high-pressure ridge which had persisted for some days—so if this were a *low*-flying balloon it was being propelled across the sky at an inordinately high speed. At high altitudes the effect of the jet stream can be considerable. If the object were at such an altitude, then it must have been of very large dimensions indeed—and frankly I do not think it was. I am familiar with weather balloons and this thing just did not fit the role. Fortuitously my wife was present at the time and was able to substantiate all I had seen.

The object continued steadily on its course and eventually disappeared over the northern tip of the island of Arran. At this point we both had the impression that it "winked out" several times before finally disappearing—in similar fashion to the object I had seen on March 1, 1938. On this occasion, however, I could not be absolutely certain due to the strong light of the setting sun and the irregular nature of the skyline. The next day some newspapers mentioned an odd aerial object performing peculiar maneuvers spotted by coastguards off the northwest coast of England. Perhaps there was a connection, perhaps there was not.

As I have emphasized already—and I make no apology for doing so again—I am not endeavoring to claim that my experiences constitute instances of UFO sightings. Certainly, so far as I am concerned, they were unidentified flying objects—inasmuch as they were flying and I was unaware of their true nature or origins. As the years pass I still find them intriguing, illogical, and peculiar. In that dark night sky of March 1, 1938, *nothing* should have been moving—yet without a doubt something *was*! I remain utterly convinced that it was no aircraft. The circumstances are such that it just could not have been. These are the facts. I have not embellished them. I cannot offer a rational explanation—and I hesitate for obvious reasons to imply extraterrestrial origins.

What I have described are, I realize, rather unspectacular events, especially in comparison to some of the claims made. However, I feel that straightforward accounts of simple events which might just have extraterrestrial connotations will do more to engender real, healthy interest than absurd, contrived accounts of flights in "flying crockery" with little green men from Venus or seductive maidens from Mars. At the same time I would certainly not wish to discredit a number of quite sincere and honest individuals who have undoubtedly during the last quarter of a century had some odd and frightening experiences.

A final little incident which, as it turned out, had nothing to do with UFOs or the like took place several years ago. It was a night in early autumn and having completed my nightly watch on certain variable stars I was currently observing I had retired indoors. I had not long been there when the telephone rang—an acquaintance of mine. He sounded breathless, almost disbelieving, and his first words were "Have you seen it?" When I inquired the nature of "it" his reply was "The thing in the sky of course, that great green light." This sounded dramatic—with traumatic undertones! However, I assured him calmly that, being comfortably ensconced in front of a TV set, I was hardly in a position to offer an opinion. Clearly, something was happening out there in the sky and I rushed outside expecting I knew not what. My informant had used the words "fairly well up in the northwest." Sure enough there *was* something there—not all that bright really but certainly quite obvious. It was roughly circular and of a very delicate pastel green. It also looked rather tenuous. I watched it for several minutes during the course of which my phone rang several more times as other friends sought aid and counsel regarding this "thing" hovering menacingly above their rooftops.

In some respects it looked to be of an auroral nature and the period being one in which auroral activity was then at a peak, this seemed the most probable answer. Nevertheless, though I have witnessed many lovely auroral displays and am reasonably familiar with the forms aurora can take, this one was certainly a departure from the book—to me at any rate! From the start I felt quite sure that the object had no connection with "UFOlogy." Now, as I watched, it moved upward toward the zenith, slowly expanding as it did so. The color grew steadily less pronounced and its substance more tenuous (by now stars could be seen shining through it). Moreover, its edge was becoming very diffused and quite clearly, whatever the nature of the thing, it was now in the act of dispersing. *That* night, at least, neither men nor maidens from the stars would be visiting us.

Press reports the following day explained the phenomenon. From an army range in the north of Scotland a sounding rocket had been fired into the upper atmosphere whereupon it had released a cloud of barium (which produces a typical green color). Mystery solved! Afterward, however, I got to thinking. Here in essence was the stuff of which flying saucer legends are made. Had there been no clear, unequivocal official explanation the next day some witnesses might have been prone to declare that a UFO had been in their sky that night. To this day I have a suspicion that my friends (who were not astronomically minded) were just a little disappointed. And frankly so was I!

I must admit that during the early years of the UFO cult I tended to have little patience with the entire business despite the personal experiences I have just recounted. To a large extent I think this was because many of the alleged "sightings" were absurd and in many cases clearly contrived for publicity purposes. It was rather evident, too, that a large number were simply examples of honest misinterpretation of natural phenomena. As the years have passed, however, and I have seen representatives of our own kind put their first faltering footsteps into space and thought more and more of races so much ahead of our own, I have gradually changed my outlook. This does not mean that I am any more willing to accept tales of little green men or antennae-eared Martians, for these things can do—and have already done—a considerable and possibly permanent disservice to what is a very intriguing and probably vital subject. But I am now fully prepared to accept the premise that there *are* alien beings, that they *could* reach here, and that if they do, events could prove exciting, unpleasant, or both. There is not a single occasion now when standing alone under a moonless, dark, star-powdered sky that I am not acutely conscious of a mildly eerie feeling. It is not unpleasant—indeed, it is pleasantly frightening if the reader will excuse this apparent contradiction in terms. At these moments I have not the faintest idea just how far removed from our world is the nearest alien intelligence, *but I have no doubt whatsoever that it exists!* No longer is it possible to stand there with the feeling that we on Earth are a populated, single, small island in a vast, lovely, but lifeless immensity. As I look at these stars I visualize a myriad other races, wonder about their histories, wonder about their future—and then inevitably about ours!

Increasingly, I feel that a number of disturbing and downright frightening experiences which a small number of people have had over the last few years are not only real but inexplicable in terms of contemporary science and technology. Increasingly, I sense that something *is* up there (not necessarily hostile or malevolent) and that it is watching us. The way our world and purported "civilizations" have behaved this is hardly perhaps very surprising.

# 14. RETURN OF THE ALIENS?

By now it is time to think of the present. The main problem here is to decide when the "present" really begins so far as the theme of this book is concerned. If one is a strict adherent of the cult of UFOlogy, then the period 1947 onward is the only possible claimant to the title. However, although the years of the flying saucer dynasty certainly cannot be omitted, it is undesirable to lay too much stress on them in view of the dubious character of many of the so-called sightings. Neither should it be our aim to denigrate this era, for without a doubt it has its allotted place in the overall fabric—perhaps a highly significant one.

It seems to the writer that the present in this context begins with the dawn of the twentieth century. Let us therefore go back a few decades, to around 1900, and look at one or two noteworthy and somewhat inexplicable events during the years between then and now. The early years of this century are, no doubt, to many a rather distant era. To many more (including the writer) they are only something from the pages of the history book. This was the era of gaslights and streetcars, or horse-drawn cabs and steam traction. But in view of the fact that we have been tracing some kind of path from Paleozoic times we can justly regard it as very recent.

Where precisely should we start? Perhaps June 30, 1908, is as good a time as any, in a place with the strange-sounding name of Tunguska.[1] Tunguska, as reference to the map will show, is in central Siberia about 500 miles to the north of Lake Baikal. What happened in this remote region that June day in 1908?

What happened was tremendous—and still largely inexplicable. The effects were massive, devastating, and far-reaching. Most readers will have heard of the Trans-Siberian railway, which straddles central Asia linking Moscow with the Pacific port of Vladivostock. On that particular day one of its great transcontinental trains had reached a point in its long journey some 400 miles to the south of the Tunguska region. It is not very difficult to imagine the scene—the flat, rather desolate country, the single line of gleaming railroad track, the long string of passenger and baggage cars hauled by a massive, powerful steam locomotive, the plume of smoke and steam streaming pennantlike behind it, the huge, churning, driving wheels. To the engineer and fireman on their pounding, swaying juggernaut this is just another day, another trip. From the footplate, hand on the regulator, the engineer watches the track ahead. The fireman, illuminated by the glow from the open firebox, steadily shovels in coal. Suddenly, inexplicably, terrifyingly, the thin metal rails, the roadbed itself just a few hundred yards ahead of the speeding express, begin to move. The rails do more than just move; they bend, buckle, and twist as if suddenly imbued with a life of their own. With a hoarse cry that is half fear, half warning the engineer whips off steam and grabs frantically for the brakes. The twisting, useless rails rush toward the train. The brakes grip and shriek. No use, too near! Disaster, complete and utter, is seconds away. With the strength and urgency born of desperation he releases the brakes and flings the great locomotive into full reverse. Showers of sparks and gouts of lurid flame leap from the massive driving wheels as they churn madly now in the opposite direction. Still, inexorably, the momentum of engine and heavy train bears them forward. But she is slowing, clawing the rails, drawing dramatically to a halt. In the coaches passengers are hurled to the floor, baggage and personal possessions cascade from the racks. With but a few feet to spare the Trans-Siberian Express comes to a sudden, dramatic, and unscheduled stop. A strange silence ensues for a second or two punctuated only by the rhythmic beat of steam from the stationary locomotive and the first shouts from excited and mystified passengers. Engineer and fireman jump down from the engine and walk a few feet up the track. What they see completely dumbfounds them. The entire roadbed has shifted and the rails are now distorted in a way that defies description. They look around. All seems at peace. There is evidence nowhere of avalanche, sudden torrent, or subsidence—absolutely nothing that could account for this alarming occurrence.

To find the reason for their close brush with death and disaster the crew of the locomotive would have had to look very much farther afield—400 miles to the north, in fact, where something from space had slammed violently into the Earth causing an explosion of monumental proportions.

The passengers and crew of the Trans-Siberian Express were not alone in the knowledge that something quite extraordinary had occurred, for the sheer violence of the impact had created seismic waves of such magnitude that they had been recorded by at least four observatories, the most distant being at Jena in central Germany. Five hours later in England, barometric readings indicated some unusual event of great magnitude. And this was by no means all, for already peculiar manifestations were appearing in the sky there, notably a remarkable red sunset glow. These effects were discussed at a meeting of the British Astronomical Association held the next month. The following is an extract from the Association's record: [2]

> About 9.30 P.M., as seen from Greenwich the sunset did not differ much from that of a normal fine summer day except that the luminosity lay more to the north. By 1.00 A.M. it had shifted a little to the east being if anything more brilliant by now. The most remarkable feature was the intensity of the light over the whole northern sky which by then resembled the southern sky under the light of a full Moon.

This effect in all probability was due to dust raised by the impact and subsequent explosion which was then borne westward around the entire globe, much as the great eruption of Krakatoa had been twenty-five years earlier.

At 7:00 P.M. on the same day we find a Mr. de Veer of Haarlem, near Amsterdam in Holland, reporting "an undulating mass to the northwest which was not cloud for the blue sky itself seem to undulate."[3] (Were these shock waves?)

It may seem odd that despite these peculiar occurrences we have to wait until 1927—nineteen years later—for the first detailed account of the Siberian event. In 1908, Imperial Russia was a troubled, backward, and uneasy land governed by an autocratic, weak, and ill-advised czar. Communications were poor and Tunguska was exceedingly remote. Certain vague reports may have filtered through to the capital, St. Petersburg (now Leningrad), but if they did, no action was taken and all record of them must have been irretrievably lost in the long years of war, revolution, and civil war so soon to follow.

In 1927 an expedition under the auspices of the USSR Academy of Sciences and led by Academician L.A. Kulik made an initial investigation of the area by air. To their surprise, the devastation, even after nineteen years, was dramatic and extensive. Over a radius of 25 miles trees had been blown down. Some had been totally uprooted and lay with their shattered roots pointed toward the center of the blast like the spokes of some fantastic and monstrous wheel.

Further expeditions by land followed during 1928 and 1929. These revealed that the region had also been severely ravaged by fire. Heat searing could be traced to a distance of from 6 to 11 miles from the supposed point of impact. Though first thoughts centered on the fall of a great meteorite no vast crater was found—only numerous small depressions none of which, despite test borings, yielded any meteoric material. This was at once odd! The famous Barringer Meteorite, which landed in Arizona before the dawn of history, left a crater 4,000 feet across and 600 feet deep.

Exploration of the region by the scientists was not easy. Contrary to popular belief the forests of Siberia are as impenetrable as the rain forests of South America, Africa, and Southeast Asia. They are also as hostile, and though the latter have the restraints of heat and high humidity the former have the devastating and numbing effects of sheer cold.

The population of the region is thinly scattered, primitive, and was at that time also highly superstitious. As a consequence it was difficult to secure active help in the shape of guides. Most of the inhabitants simply refused to venture near the actual impact point of the object, saying it was where "the great God Ogdi had come down from Heaven to Earth and destroyed by fire all those who entered there." Kulik and his men had to traverse the region on their own bearing a number of heavy items such as drills to penetrate frozen ground. The total absence of roads, even cart tracks, did not exactly facilitate matters.

But if the inhabitants of the region were reticent about offering practical assistance, there was one way at least in which they were ready enough to help out. Very poignant and dramatic descriptions were made to Kulik and his colleagues by a considerable number of those who had been around that fateful and terrifying morning.

Said Vassili Ilich, once the proud owner of a fine herd of five hundred reindeer, of the object's descent: "The fire came by and destroyed the forest, the reindeer and all other animals."[4] This was perfectly true. Shortly after the blast the hideously distorted remains of a few reindeer had been discovered. Of the rest there was no trace whatsoever. They had simply been vaporized out of existence! At Vanovara, some 40 miles to the east, a farmer named Lemenov spoke of "an enormous fireball in the north and a fierce terrible wave of heat which burned the shirt from my body." [5] Just for good measure the unfortunate man was also picked up and hurled for several feet, his barn wrecked, and every pane of glass in his house shattered. Fellow farmer and neighbor P. P. Kosolopov said it was as if "the ears were being burnt from my head." [6] He sought refuge in his house, which promptly collapsed about him. Another very peculiar manifestation was the fact that Vassili Ilich's silverware and tin utensils are said to have *melted,* though some large buckets (material unknown) did somehow manage to survive intact. According to other witnesses, the "object" was *seen* in a cloudless sky over an area roughly 1,000 miles in diameter and was likened to the flight of a blindingly bright bolide "which made even the light of the Sun seem dark." [7]

Reports from countries other than England and Holland told of the peculiarly bright nights that followed. From all accounts most of Europe and western Siberia were affected in this fashion.[8] Indeed, as far south as the Caucasus Mountains it was possible to read a newspaper at midnight without artificial light. This peculiar luminescence disappeared gradually over the ensuing two months. Other effects of a less obvious nature were also observed. The observatory at Irkutsk, among others, reported, for example, a definite disturbance in the Earth's magnetic field.[9]

According to E. L. Krinov, the dazzling "fireball" moved within the space of a few seconds from southeast to northwest leaving a trail of what he described as "dust." [10] Flames and a vast pillar of smoke were observed over the immediate area of the occurrence.

Indeed, spectacular visible phenomena were seen from as far away as 400 to 500 miles, loud detonations being heard afterward at distances in excess of 600 miles. In the light of these facts one can readily appreciate the cataclysmic effects the event would have produced had it occurred in a densely populated area. Had it taken place in the United States, say in the vicinity of Chicago, visible phenomena would, it is reckoned, have been observed as far to the southeast as Pittsburgh, as far south as Nashville, and as far west as Kansas City. Its dreadful thunder would have been clearly audible in Washington, D.C., Georgia, Oklahoma and North Dakota (Figure 24).

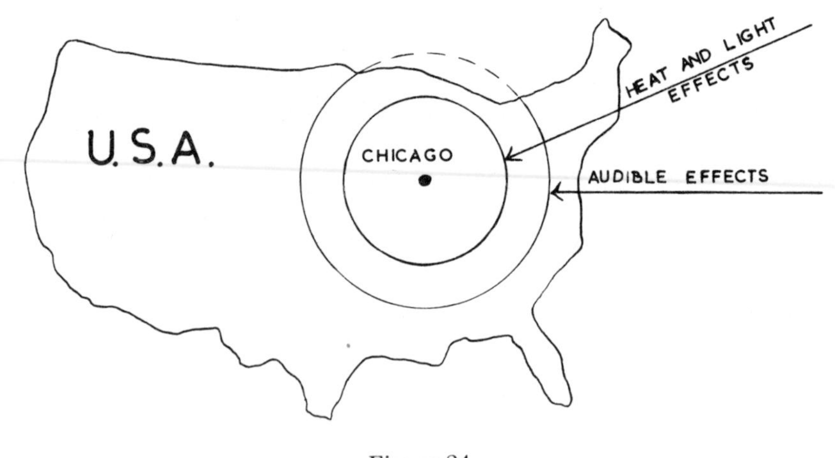


Figure 24

Obvious first impressions suggest that the thing must have been a meteorite of mammoth proportions, but we have seen that it left no crater on the Arizona pattern. One explanation for this apparent anomaly is that the "meteorite" simply destroyed (i.e., vaporized) itself by the sheer violence of the explosion. The weakness of this argument is the absence of anything resembling meteoric debris and the lack of even small craters. It seems incredible that so vast a mass of rock and/or iron could have vaporized itself entirely. On the basis of the resultant earth tremors the mass of the meteorite—if this is what it was—must have been in the region of *10 million tons.* Certainly the valley wherein "it" fell is in the permafrost region of Siberia. This has given rise to the belief that by 1927 mud flows would have filled up the crater leaving little or nothing to be seen. It is rather tragic from a scientific point of view that no attempt was made to investigate in the days and weeks that followed the occurrence. Surely, though, the evidence of some sort of crater should have remained, if only a large shallow irregular depression. This lack of cratering is oddly significant and does little to commend the idea of a simple meteorite of large dimensions.

Next comes the concept of collision with a cometary head (presumably a comet that no astronomer or observatory had earlier spotted). This theory is due in large measure to Astapovich and Whipple [11] (working independently in 1930). Admittedly, there are points in its favor.

1. It is thought that cometary heads are largely ice, methane, and ammonia with *traces* of mineral matter. In this way the absence of solid material on the ground near the blast center might be explained.
2. The spectacular luminescence of the night sky immediately after the event (observed in Siberia, European Russia, and western Europe but not in the United States, Canada, or the southern hemisphere).

The tail of a comet is always directed away from the sun (a repulsion effect). Was the tail of this particular comet (assuming, of course, that it *was* a comet) at the moment of impact streaming outward in a northwesterly direction? Estimates made in 1961 indicate that dissipation of such a tail would initially have brightened the night sky some 50 to 100 times.[12]

The basic weakness of the comet hypothesis is the improbability that a comet, even a small one, on a collision course with Earth could have gone so completely unobserved.

It is claimed that the course of the object indicates an arrival from a part of the sky within the constellation Cetus, the Whale. (It did not, of course, actually *originate* within the stars of that constellation.) V. G. Fesenkov has estimated that the object, whatever it was, had a diameter of several hundred meters![13,14]

As far as our theme is concerned we now come to the really

interesting part. A. Zolotov has suggested that the energy expended as the object broke up was *nuclear*. His theory is based on the following observations and premises.[15]

At a distance of approximately 10 to 12 miles from the apparent epicenter he found trees, subjected to thermal flash at the time of the blast, which had started to burn. A normal forest fire can be ruled out since a living tree requires 60 to 100 calories per square centimeter of incident thermal radiation to start ignition. Calculation places the radiant energy of the explosion at $1.5 \times 10^{23}$ ergs. This is in quite close accord with an earlier calculation based on the direction of the body's flight being opposed to that of Earth. The resultant high collision velocity (60 kilometers per second) would yield an impact energy of $1.0 \times 10^{23}$ ergs. Thus the estimated yield of thermal energy is due to that of the total explosive energy. This, claims Zolotov, points to a nuclear rather than a physical explosion.

The burning sensations experienced by farmers Semenov and Kosolopov (who were 40 miles distant at Vanavara) and the melting of Ilich's metalware all lend a measure of support to this hypothesis. For the record, silver melts at 960.5°C and tin at 231.9°C. Thermal radiation reaching nearly 1,000°C at a distance of 40 miles is a considerable performance to say the least. It would seem that only a nuclear explosion could adequately explain it. A fraction of a second after such an explosion a high-pressure, intensely hot, and luminous shock front moves outward from the initial fireball.

Recently the suggestion that "anti-matter" was the agent involved has been made. This is rather a complex concept. Simplifying, we can use the hydrogen atom as an example; it has a positively charged nucleus around which orbits a single negatively charged electron. The opposing charges cancel each other out so that the atom is rendered electrically neutral. In the "anti-hydrogen" atom (assuming such a thing exists) the converse applies, that is, there should be a *negatively* charged nucleus and a single *positively* charged electron (Figure 25).

Physicists believe that matter and anti-matter brought together would be mutually destructive—and in a most violent fashion. Could the Tunguska object have been a chunk of anti-matter? If by chance it were (and the odds against the possibility do seem

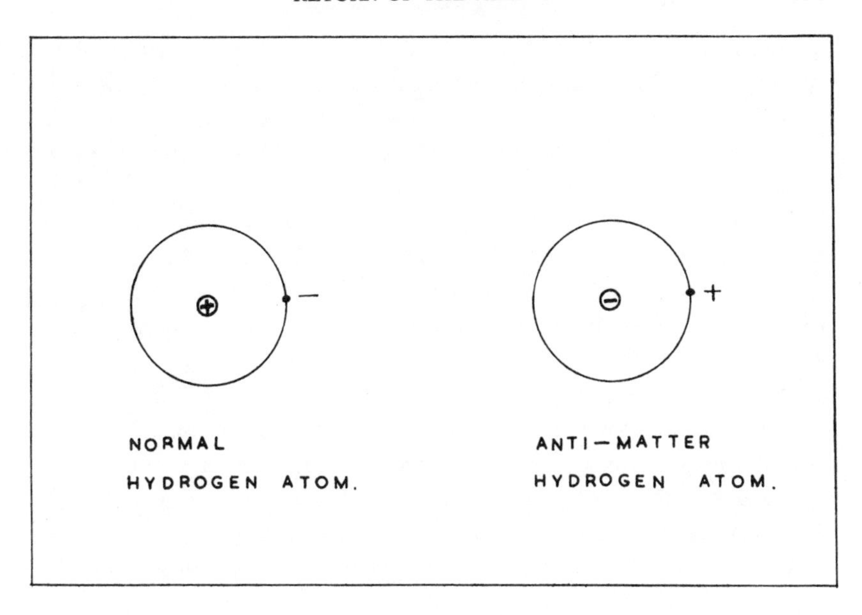


Figure 25

long), it could hardly have originated in the solar system and would therefore have to be regarded as an intruder from somewhere in interstellar space.

In some respects the anti-matter hypothesis has a mantle of credibility, for there is little doubt, in view of all the evidence, that the "thing" disintegrated violently, creating shock and searing heat waves of an intense kind. It seems unlikely however that large chunks of anti-matter are floating about loose within the confines of the solar system or, for that matter, in interstellar space. For this reason alone it should be treated with the greatest possible caution. Doubts based on more precise reasons have recently been expressed by Venogradov and several other Soviet geochemists. Proponents of the anti-matter hypothesis have stressed not only the great amount of energy released by the explosion but also the marked increase (7 percent) in the amount of radioactive carbon in the atmosphere. Venogradov, however, measured the quantity of carbon near the epicenter of the explosion by the annular rings in the trunk of a 140-year-old larch tree. He found there was indeed an increase in the radioactive carbon content of the atmosphere, but that in 1908 this amounted to only 2 percent near the epicenter of the blast.

It has been further demonstrated by research workers in several countries that variations in the carbon content of the atmosphere have on occasions during the past three thousand years amounted to as much as 4 percent. These are believed to be related to solar activity. Had anti-matter really entered the atmosphere that June morning in 1908 the figure would almost certainly have been in excess of 2 percent.

We have seen that a "meteoric" explanation, though seeming the most plausible, has certain inherent weaknesses. No great crater was produced (from all accounts not even minor ones), thermal effects were greater than would normally be anticipated, and certain aspects of the situation would seem to have nuclear connotations. Of considerable interest also in this respect are the words of witnesses who claim to have *seen* the "thing" as it hurtled toward the ground. Especially significant are those accounts which describe it as a "bright blue tube."[16, 17] No meteor has ever looked like a bright blue tube—or is ever likely to! What, then, was the nature of the thing that these men claim to have seen? Naturally, one must have certain reservations about descriptions by peasant farmers. Excitement and fear can have peculiar repercussions, especially with respect to personal recollections. It is essential also to recall that nineteen years had passed since the event. Details become easily embellished with the passage of time. In this way is legend born. And what one person saw (or thought he saw), so by suggestion will others. Nevertheless, the reports are intriguing and cannot be arbitrarily dismissed. If what these men said is in accord with fact, then the event takes on an entirely different hue.

We have also seen that the "anti-matter" idea, though novel and interesting, is not very likely to hold up. Where, then, does that leave us?

A Russian scientist, A. Kasantzev, first gave expression to the most radical explanation of all. Kasantzev had, along with other scientists, visited the ruins of the Japanese city Hiroshima not long after its destruction in 1945 by the atomic bomb. There was much to see, much to examine, much to ponder over. The sight remained vivid in his mind. It so happened that a few years later he was also in a party visiting the sight of the 1908 Tunguska blast. Forty years had now passed since that fateful day in June 1908. The effects were much less stark but the scars and devastation were still there. As he moved around something began to gnaw in his mind. There was a certain familiarity about much of this, especially the searing and the burning. Suddenly it came back to him—Hiroshima! Things he was seeing here he had seen before in the ruins of that luckless Japanese city.

There was, he observed, a particularly marked similarity in the searing effects on vegetation and in the form of the blast damage. An atomic device had been exploded deliberately above the city of Hiroshima. Here it was as if an atomic device had been exploded above the area too—presumably without deliberate intent. In 1908 there were no atomic bombs or any other nuclear devices. The more Kasantzev thought about it, the more convinced did he become. The Tunguska "thing" had been *nuclear.* It could not have originated on Earth so therefore it must have done so somewhere beyond Earth. The other worlds of the solar system could almost certainly be ruled out. Therefore, it must have come from *beyond.* It had to have been a starship and since it is unlikely that its occupants intended to immolate themselves it must have exploded of its own volition, perhaps as it tried unsuccessfully to effect a landing.

This was a daring theory, especially at the time. To a certain extent it still is. Yet some aspects do add up—possible nuclear origins, reports of an object being *seen,* the violent and far-reaching effects, the description of a colossal fireball, the terrible wave of heat, the melting of silver utensils. Incidentally, a recent calculation indicates that if this *were* a nuclear detonation it must have been in the 0.2- to 20-megaton range! [18]

Now no one should be so rash as to claim categorically that this is the case of an exploding starship. The idea can be presented only as a possibility. Without doubt the blast was of monumental proportions. If meteoric in origin, then it is unlike other known meteoric falls and shows few orthodox meteorite effects. If it was the head of a small comet, then how did such an object get as close as this to Earth without being spotted? Every year several faint small comets are discovered and tracked by astronomers, as often as not by amateurs. Enthusiasm in 1908 for this form of sky-watching was very great. It is almost inconceivable that the approach of a comet which subsequently collided with Earth could have escaped unnoticed. As far as the "anti-matter" idea is concerned, it is advisable to keep a tight hold on the reins since this is so speculative.

Fairly recently the affair was attributed to collision with (of all things!) a small Black Hole.[19] This is assumed to have had the mass of a small asteroid and a velocity slightly greater than the escape velocity of Earth (7 miles per second). The originators of this idea believe that most of the radiation from the shock front would lie in the ultraviolet portion of the spectrum and would be reabsorbed and reradiated at longer wave lengths. Consequently the accompanying plasma column would appear *deep blue.* This is certainly relevant, for we have already seen that certain witnesses spoke of a "bright blue tube." Additionally, a Black Hole would leave no crater or material residue—both points in favor of such a theory. It would enter the Earth, and in view of its high velocity and because of the fact that it would lose only a small fraction of its energy in passing through the Earth, it would pass straight through the planet, entering at 30 degrees to the horizon and leaving through the North Atlantic Ocean in the region 40°–50°N, 30°–40°W. The originators believe that at the exit point there would be another air shock wave and disturbance of the sea surface. They suggest that microbarograph records should be checked for an event similar to that caused by the entry shock, and that oceanographic and shipping records be studied to see if any surface or underwater disturbances were observed. The postulated area of exit is centered on a point midway along a line joining New York and London.

The odds are very much against such a radical theory.[20] It is clear from the extent of the damage—trees blown down over a radius of 30 to 40 kilometers and heat-seared within a radius of 15 to 18 kilometers—that most of the energy was dissipated in an explosion. The exposed roots were also directed toward the center of the area, and in ravines partially protected trees remained standing, though many had their tops broken. In the circumstances the explosion would appear to have taken place *in the air.* A conventional solid nickel-iron meteorite would, on the other hand, have first buried itself below the surface then spent its energy in a subsurface explosion. An air blast *would* almost certainly result from the impact of a small Black Hole of asteroidal mass. Microbarograph records have not, however, revealed any sign of waves from the suggested exit explosion in the North Atlantic. Neither have shipping records. In view of the nature of Black Holes one would also feel that, cataclysmic though the blast undoubtedly was, collision with a small Black Hole would have been infinitely worse—at the point of exit as well as entry. We might even wonder whether our planet could be expected to survive such an experience. Eyewitnesses spoke also of a thick dust train along the path of the object immediately following its passage. This is considered consistent with the deposition of material in the atmosphere rather than loss of air into a Black Hole.

We can, it seems, eliminate anti-matter, Black Holes, and conventional meteorites. This leaves us with the cometary head collision and the exploding nuclear spaceship hypotheses. There is much to commend the former, especially the brilliance of the succeeding nights, the air explosion, the absence of a crater or craters. Less easy to explain is the degree of heat searing and the fact that a comet, even a small one, on a collision course with our planet had not been spotted a few hours beforehand.

Only, it would appear, could the Black Hole collision concept explain the *natural* appearance of a bright blue tube. Several observers spoke of this phenomenon, which means at least that it *may* have been real. Where then does this leave us? A space vehicle hurtling earthward out of control might give such an impression. That does not seem unreasonable. It might also be added that numerous small magnetic globules with a high nickel content were afterward found in the Tunguska region. Were they part of a comet's nucleus (head) or the fused remains of "something else"? Silicate-magnetite spherules were also found near the Tunguska epicenter.[21] No such cosmic remnant has ever been found anywhere else in the world.

Kasantzev closed his account with a poignant observation and in the circumstances it seems only fitting that it should be recorded here. "On that day," he wrote, "visitors to our world, creatures of unknown form, may have perished. On a world far remote from our own, memories may still survive of loved ones who set out but were destined never to return."

We will now jump in time nearly five years. This time our interest centers on the other side of the globe, though still in the northern hemisphere. On February 9, 1913, many people in Canada were startled to see a procession of what were described as "glowing red fireballs" moving slowly and majestically across the sky, apparently in level flight. Their altitude was estimated at 50 miles and the sound of their passage likened to the "roar of an express train." Now this, to say the least, is a very odd way for meteors ("shooting stars") to behave. Most of us have seen the occasional bright meteor—a sudden, silent, darting streak of light across the starry sky. Only if one falls to Earth are there sound effects. Meteors also occur in showers, the best and most prolific of these generally occurring in the autumn or early winter. In the past some really marvelous displays have been recorded, for example, the fabulous 1866 display of the Leonids, so named because they appear to emanate from a point or radiant in the constellation Leo, the Lion. Many of these meteor showers have degenerated with the passage of the years so that what we are more likely to see now are several bright meteors appearing within a few minutes of one another rather than a great host at once.

The event of February 9, 1913, could not be construed as a meteor shower, nor even as a group of isolated meteors, for these bodies do not pass across the sky in level flight making the sound of an express train. In the circumstances readers may be tempted to inquire whether this is just an example of a normal event that has become increasingly enhanced with the passing years. In fact, the phenomenon was investigated at the time by Professor C. A. Chant, of the University of Toronto, who verified that the phenomenon had been sighted from Saskatchewan in western Canada, across the Great Lakes to the eastern coast of Canada and the United States. Indeed, subsequent reports were also received from points as distant as Bermuda and Brazil. In each and every case observers stressed the apparently level flight and "rushing noise" of the so-called meteors.

This is really a very odd celestial manifestation. If these were in fact meteors, then they were certainly behaving in a peculiar manner. As one who has been watching the skies fairly constantly since 1938 and in that time seen innumerable meteors the writer can truthfully say that never on any occasion has he seen anything remotely resembling this.

A more recent inquiry into this episode has suggested the possibility that what the many observers saw that February night in 1913 was the spectacular demise of a small and hitherto unsuspected natural satellite of the Earth (a second moon), the main mass of which subsequently plunged into the waters of the Atlantic Ocean. This presumably is feasible, however remotely. Perhaps something did fall into the Atlantic, but was it a tiny moon of Earth, a large meteor breaking up, or something more significant? Again, as with the Tunguska "object," we cannot in all conscience claim this was an alien space vessel in dire trouble. To do so would require the tangible evidence of the remains, assuming anything recognizable were left. Space vehicles in dire peril are not likely to leave clear evidence of their existence. They explode or vaporize—probably both—and that is that. Skeptics might say "how very convenient"—and in a way who could blame them? If something did plunge into the waters of the Atlantic, then, presumably, in some form or another, it, or bits of it, are still there, state unknown. Location of items on the ocean bed is never an easy undertaking even when the precise point of entry is known. In April 1912, off the south coast of Newfoundland, the greatest and most prestigious ocean liner of its day, the mighty 45,000-ton *Titanic,* foundered after colliding with an iceberg. Seventy-three years were to elapse before its massive shattered hulk was located, filmed, and televised as the result of a painstaking search by Dr. Robert Goddard and his team from the Woods Hole Oceanographic Institute, Massachusetts.

Presumably, if an event similar to that of February 9, 1913, were to occur today it would simply be labeled "meteorite" and left at that. Only if the object plunging into the sea were clearly seen to be some form of aerial craft could there be justification for a search and a serious attempt at recovery.

While on the subject of objects plunging into the sea we should also mention an event said to have occurred in August of the same year. Unfortunately, documentation of this event is lacking and it must therefore be included in these pages with considerable reserve. The facts of the situation, for what they are worth, are these. While sailing near the coast of Java, in what was then the

Netherlands East Indies, the crew of a small British freighter were astonished to see a gigantic "wheel" come spiraling down from the sky and plunge into the sea. Although the ship's master and several others saw it for only a few seconds, all claimed that it was of vast dimensions, gleamed strongly in the bright tropical sunlight, and that it was embellished by strange designs and hieroglyphics. This was well before the age of flying saucers and UFOs. On the other hand, sea lore is full of strange tales and events. Perhaps this great "aerial wheel" belongs to the world of the *Marie Celeste,* the sea serpent, the "Flying Dutchman," and maps inscribed "here be dragons." Only this *was* the twentieth century!

There is also a most interesting account, published in 1916 by Walter Maunder, an astronomer at the Greenwich Observatory, London. The actual event, however, had taken place thirty-four years earlier, on November 17, 1882. We are admittedly going back beyond 1900 but since this particular event took place in the closing decades of the last century we may be forgiven this breach of our earlier self-imposed restriction. Moreover, this is a report by a qualified professional astronomer. Such people are trained observers and among the least likely to disseminate ridiculous tales. Perhaps we should allow Mr. Maunder to describe the circumstances in his own words:

> A great circular disc of greenish light suddenly appeared down in the east-north-east as though it had just risen, and moved across the sky, as smoothly and steadily as the Sun, Moon, stars and planets move, but nearly a thousand times as quickly. The circularity of its shape was merely the effect of fore-shortening, for as it moved it lengthened out, and when it crossed the meridian and passed just above the Moon its form was almost that of a very elongated eclipse, and various observers spoke of it as "cigar-shaped" or "like a torpedo."

It should be added that the time was just after sunset and Maunder was looking across London from the roof of the Greenwich Observatory, which lies to the southeast of the city.

Maunder was not the only person to witness the strange phenomenon, for it was seen by hundreds of people throughout Britain as well as on the mainland of Europe. So many sightings came in, some, like that of Maunder, of such impeccable pedigree and accuracy, that it was possible to obtain some estimate of the height, dimensions, and speed of the peculiar apparition. These showed that it was 100 miles above the surface of the Earth, was moving at approximately 10 miles per second (3,600 miles per hour), and was at least *50 miles* long!

The obvious conclusion is that this was probably auroral in origin but certainly of a rather unusual type. Records for that day, and for some days preceding it, show that a fairly violent magnetic storm was then in progress. In these circumstances we really have to settle for a natural explanation. The only jarring feature is the odd configuration of the phenomenon, which tends to put it outside the normal gamut of auroral forms.

One particularly curious feature which characterized the early years of the present century was the occurrence from time to time of certain inexplicable "booming" sounds. Indeed, this had already started toward the close of the nineteenth century and has, in fact, been reported in isolation during recent times. Now there could be many rational explanations for such sounds. The obvious one nowadays would be sonic booms but this reason is clearly not valid with respect to the early years of this century and those preceding them. Those were of course the years when navies with large gun battleships were coming into existence. The sounds from these huge rifled cannons can carry over large distances. Then of course we must also take into account the possibility of distant thunder or blasting operations.

As it turns out, the position with regard to these sounds is not as simple as all that. In many instances, the authorities flatly denied the firing of large naval guns or the existence of blasting operations and meteorological reports from the regions concerned showed there had been no electrical storms to produce thunder. From whence then came the sounds?

Records show that they were heard in Britain, in Iceland, off the Belgian coast, in parts of the United States (Montana, Wyoming, and Dakota), in parts of Siberia, in Haiti, and in Australia. The words of H. L. Richardson of Caernarvon, Western Australia, on June 30, 1908, are especially interesting: "I heard three explosions in the air, at a *great height,* followed by a *noise like an escape of steam,* lasting several seconds." Now, more than eighty years later, it is clearly impossible to get amplification of this report. This is unfortunate because it is extremely interesting in two respects—the explosions were allegedly heard *"in the air at a great height"* and were followed by a noise like *"an escape of steam."* It is almost as if one of the great naval dreadnoughts of the time had taken to the air, then burst its boilers! Could it have been an accident at sea with the sound waves refracted in a way which made them seem to come from above? Rather unlikely—and anyway no such mishaps were reported from anywhere in the region at that time. (Could there have been a connection here with the Tunguska "object"? The date coincides exactly!) Very tempting at this stage, of course, to visualize a space vessel a few miles up in dire trouble—only in those days there were no space vessels—*as far as we know!*

Another rational possibility with respect to distant reverberating sounds is that of large-scale volcanic activity. When in August 1883 the island volcano Krakatoa in the Straits of Sunda between Java and Sumatra blew itself cheerfully into a volcanic Valhalla the sound was heard, after appropriate time intervals, in places hundreds, even thousands of miles distant. The official record is 4,000 miles, the rumbles having been heard in Rodriguez Island in the Indian Ocean. Statistics show, however, that in nearly all instances the inexplicable celestial rumbles did not coincide with volcanic eruptions even allowing for time lag.

What then is the answer? A natural phenomenon still not understood, naval firing practice despite official denials, a form of self-delusion? The first seems unlikely in the light of today's knowledge; if the second, then the world's navies seem to have had plenty spare ammunition to expend; and as for the third, well, rather a lot of people must have been deluding themselves. There, regrettably, in the absence of further information, we must leave the subject.

During the early years of this century one is struck by the large number of highly spectacular meteorites reported. In fact this begins in the 1850s. In preparation for this book the writer scrutinized all volumes of the noted British scientific weekly *Nature,* which began publication in the 1880s. Now it is certainly not our intention—and this cannot be too strongly emphasized—to suggest that the many reports all relate to manifestations of alien pyrotechnics. Most, the vast majority in fact, are clearly meteoric in origin. Of this there can be no doubt whatsoever. As we enter the 1920s, however, the incidence of these reports declines noticeably. Now this could be a decline in the number of meteors or just simply a decline in the number of reports. Around the turn of the century, however, and for a few years in either direction there occurred a few meteors which can only be described as highly spectacular to the point of being peculiar. Since 1938 the writer has been observing the heavens assiduously and regularly. He has seen many thousands of meteors. Most have been utterly *un*spectacular; roughly 10 percent have been bright, very beautiful, but *always* typically meteoric. At no time has he ever suspected that the object was not what it appeared to be. He cannot say the same for the turn-of-the-century reports. Of course, these reports could have been embellished. That possibility must be faced, but even allowing for embellishment, some of these meteors did perform in odd ways. Probably there is nothing in it—yet in the mind of the writer at least, there has always remained that slightest of slight suspicions.

From the 1920s onward, meteorites apart, nothing very much of an unusual nature seems to have occurred in the skies over this world. The late twenties and entire thirties are largely barren. Had people too much else to think about? Did economic slump and approaching war simply bring down the curtains on a celestial "silly season" or was there some more significant reason?

In the mid-forties strange phenomena began again. Was this something totally new brought about by man's acquisition of dangerous knowledge or simply a further return in greater strength? If the latter, time in the end will surely provide the proof!

# 15. UFOS AND ALL THAT

And now at last we come to the contemporary scene. Like it or not, this means for the most part UFOs, "flying saucers," and all manner of other peculiar (and ofttimes improbable) space- and air-borne crockery. At the outset let us be perfectly frank and admit that there has been more rubbish spoken and printed on this subject than about any other in modern times.

Nevertheless, the peculiar reports of the last thirty years are intriguing, for nothing quite like this has ever happened before. Many of these alleged sightings have been reported by perfectly sincere, honest, and well-meaning persons whose intention was neither to defraud nor secure a dubious fame. As we will seek to show, however, the eye can at times be rather easily fooled, especially if it happens to be untrained. For this reason many of these reports can almost certainly be attributed to misinterpretation of physical and meteorological phenomena. These people are not to be blamed. Culpability lies elsewhere. In any event the scene had been well set for them and their minds in a sense preconditioned.

Without doubt, some sightings *are* the product of publicity seekers, hoaxers, and the lunatic fringe. We have even been treated to "photographs" of flying saucers. Some of these bear an unhappy and most marked resemblance to pan lids, light shades, and the like, and for this reason can be dismissed at once. Others are much more subtle. A faked photograph well done can be so convincing that even experts find it hard to detect the flaws. As a matter of interest, the writer has in his possession a 35mm color slide taken personally which purports to show a jetliner (a French Caravelle) making a "wheels-down" landing approach. The picture looks extremely authentic but in fact the plane is a polystyrene scale model 18 inches long, the gossamer-fine nylon thread suspending it from an unseen rope being completely invisible. So far this particular photographer has refrained from any attempted simulation of a "flying plate" on the final stages of its journey from Alpha Centauri!

A genuine error is very easy to make. This was brought home to the writer quite vividly during a holiday visit to the United States in July 1972. The locality was near St. Ignace in the Upper Peninsula of Michigan just across the Straits of Mackinac. The time was early evening with the sun beginning its dip toward the western horizon. The day had been very hot and it was pleasant now to relax under the pine trees beside that little log cabin with the waters of Lake Michigan lapping gently a few feet away. Suddenly a sinister torpedo-shaped object appeared in the sky. Seconds later the sound of a high-flying jet became audible. Due to the great disparity between the speed of light and that of sound the latter appeared to emanate from a point a considerable distance from the mysterious object. The overall effect was that of a normal invisible aircraft overhead and a sinister-looking alien craft well to the west. The mystery was soon solved however. After about a minute the "torpedo" began to change course and as it did so appeared to sprout wings and a tail fin. The strange craft was not after all the vanguard of an invasion fleet from Vega but the property of the United States Air Force going about its lawful business. At the angle at which it first appeared its wings were seen edge-on, which, at that range and altitude, rendered them indistinguishable, while the tail fin somehow blended with the sky. In contrast the fuselage of the plane looked very dark. It was this part which constituted the "torpedo." Had this aircraft disappeared into cloud before changing course or had it continued to proceed on its original bearing, an observer on the ground might easily have been tempted to report the sighting of a UFO—and that in all sincerity.

Another personal instance illustrates how mistaken identity of normal events can occur. This one has quite definite "flying saucer" connotations. Flying from England to Florida in July 1971 to observe the launch of Apollo 15 at Cape Kennedy, the Boeing 707 on which the writer was aboard touched down at Bermuda for a brief refueling stop. By now it was late afternoon, local time. The humidity was extremely high, as was the temperature. There was also a sullen sultriness which did not augur too well for a smooth continuation of the flight. Sure enough, the remainder of the journey to Miami was through weather which became increasingly thundery and turbulent. It was fascinating to watch the thunderheads form just above the surface of the ocean then ascend slowly to our altitude like great billowing clouds of cotton. From the top of one such cloud, driven no doubt by the prevailing wind at that altitude, there separated five small tufts of cloud each of which at once adopted a nearly circular configuration and proceeded on its own in line ahead. The "leader" (the first breakaway portion) was the largest. Once all were clear of the parent cloud they assumed an almost textbook "flying saucer" formation—"mother ship" and four attendant "scouts."

Had one not seen the actual "birth" of these things it would have been easy indeed to imagine they were genuine "flying saucers." Said a young stewardess whose attention had been drawn to the spectacle, "I see it—but I don't believe it!"

The writer can also testify to the ease with which a bright planet seen from the air could easily be mistaken for a UFO. Realization of this occurred during the return flight from Moscow in November 1973 following a visit to the Soviet space exhibition in that city. The sun had already set. Altitude was around 35,000 feet in a clear and cloud-free sky. Far below lay the former Baltic state of Latvia. The distant horizon had an exquisite bluish tinge which contrasted beautifully with the gold of the afterglow. There in that afterglow hung the planet Venus. At least, from the ground it would have hung there apparently motionless. From the air the position was rather different. At that particular time our aircraft was making a number of changes in course and since there was no fixed object to use as a reference point it was as if the planet were gyrating across the sky while aircraft and occupants hung suspended in the sky watching it.

Mistaken identity can also result from a feature known to meteorologists and others as noctilucent clouds. These are not uncommon and are generally attributed to dust particles at a height of about 80 kilometers (50 miles) "drifting" at speeds of roughly 640 kilometers per hour (400 mph). They are termed noctilucent because they are visible only against a night sky. The type of configuration varies but circular and elliptical shapes are among the most common. The rest follows! There is also a very powerful psychological factor at work, night being the time which generally renders the seemingly inexplicable more mysterious. With perhaps one or two stars or even a planet showing, thoughts of UFOs are readily engendered.

We can see how, with the best will in the world, it is so very easy to be mistaken. Perhaps if the flying saucer legend had never been born people might see these things and think nothing of them. But, as was mentioned earlier, people have become conditioned to such an extent that it is now almost a reflex action. A circular object is seen in the sky and at once the words "flying saucer" echo down the corridors of the mind—and quite often a few hours later down the corridors of the newspaper offices too!

However, the past few decades have produced so many reports of queer objects and manifestations in the skies (night and day) that the matter cannot in all fairness be arbitrarily dismissed as mistaken identity, hysteria, imagination, or the buffoonery of cranks. In many instances the observers have been trained, competent, and trustworthy members of society—airline pilots and crews, Air Force personnel, police, and, yes—even scientists!

The whole flying saucer/UFO business started just after the close of World War II—or, perhaps significantly, just after mankind had developed a crude capability both in nuclear fission and in long-range rockets.

It may be of interest at this point to include a breakdown of the 808 officially investigated British sightings up to the close of 1967.[1]

| | | | |
|---|---|---|---|
| 1. | Satellites and associated debris | 211 | |
| 2. | Weather balloons | 92 | |
| 3. | Orthodox celestial bodies, e.g., planets | 60 | |
| 4. | Meteorological and other natural phenomena | 72 | |
| 5. | Aircraft of all types | 226 | |
| 6. | Miscellaneous (cloud reflections, hoaxes) | 63 | |
| 7. | Unexplained | 84 | (10%) |
| | | 808 | |

Thus, while almost 90 percent of the sightings investigated could be adequately explained, over 10 percent could not. In the circumstances this is hardly a small proportion though no doubt some of the 84 could have been placed in other categories had more definite information been available.

UFOs have now been seen in more than fifty countries, the highest proportion being in the United States. Up to 1967, 11,000 sightings had been investigated by the American authorities. The writer has not so far been able to secure the breakdown figures with respect to these. If arbitrarily we accept the same proportion as "unexplained" (around 10 percent) then 1,100 sightings must go into this category. Suppose, too, for the sake of argument we prune the figure drastically both with respect to American and to British sightings to a mere tenth. This leaves a hard core of 110 in the United States and 8 in Britain—say, 120 "inexplicables" in the two countries over twenty years (1947-1967), which works out at 6 per year.

The official verdict of the U.S. Air Force in 1967 was brief, although in the absence of any really positive information, perfectly fair. It ran to a mere eighteen words and was as follows:

> No evidence has been received nor discovered which proves the existence and intraspace mobility of extra-terrestrial life.

The findings of the British Ministry of Defence were couched in very similar terms. It should be noted, however, that neither country in its findings actually poured scorn on the idea. They merely stated the facts as they then found them. What they produced was a perfect example of an open-ended verdict.

During 1967 there appeared in the official magazine of the U.S. Air Force a few paragraphs which neatly summed up all that was known about UFOs up to that time. It is well worth quoting here:

> U.F.O.'s resemble cigars, propellers, hats, pie-pans, saucers and balls. Sizes vary from tiny objects estimated in inches to massive space-ships allegedly 250 feet or more in diameter.
>
> The predominant colors are red, green, blue and white; however the entire color spectrum has been reported at one

> time or another. U.F.O. sounds range from eerie silence to high-pitched penetrating tones. Propulsion may be evidenced by flaming exhausts or there may be no exhausts whatsoever. U.F.O.'s hover, zig-zag, move in any direction at variable speeds. Structures fluctuate from solid to almost invisible.
>
> Although U.F.O.'s are observed throughout the year, the largest number of sightings occur at night during the spring and summer when people spend more time out of doors.

The article then proceeded to enumerate the causes for UFO sightings, attributing these in many instances to meteorological balloons, artificial satellites, aircraft of unconventional design, aircraft and unusual weather conditions, aircraft having unusual external light patterns, flocks of birds, reflection of searchlights or automobile headlights from low clouds, reflection of sunlight from shiny surfaces, optical mirages, peculiar cloud formations, ball lightning, meteors, planets, bright stars, and the Aurora Borealis.

There have been two full-scale official investigations into UFOs in the United States although both understandably proved inconclusive. Frequent dark reports that the authorities in America, Britain, and in other countries are aware of a real threat and are covering this up seem on all the evidence available to be quite unjustified.

There is undoubtedly a tendency now among orthodox scientists to look more closely at the subject and not just dismiss it out of hand. In this respect it is worth quoting the words of Dr. James McDonald, professor of meteorology at the University of Arizona: "Unidentified flying objects represent one of the great scientific problems of our time and the possibility that they originate beyond the Earth must be given serious scientific attention."

Why, it is often asked, if these are space vehicles carrying supra-intelligent beings from remote planets do they not come down openly and let us have a look at them? (It is not hard to imagine the panic that would ensue if they did!) The answer could easily be a very simple one. Supra-intelligent beings, just because they *are* supra-intelligent, would not necessarily take this point of view. To them it might constitute an idea bordering on the moronic! Their purpose, for the present at least, might just be to observe our planet and the ways of its occupants. If the intention is to observe the habits of a colony of ants one does not normally commence operations by planting a great foot slap in the middle of that colony. Something a little more subtle is called for. But should that colony seem dangerous or potentially so, the foot can then come down. This is a line of thought we must enlarge upon later. It represents one of the more disturbing ones.

Moreover, if a civilization, or at least its representatives, have been able to conquer space and time to reach this corner of the galaxy it is very unlikely that their spacecraft are going to be in any way like ours. An alien ship is going to be a very sophisticated craft—hence, perhaps, the "will-of-the-wisp" potential with which they seem to be endowed. Such a craft would almost certainly bear little semblance to a space vehicle as we understand the term. Indeed, it has been suggested that such a craft could have been rendered invisible or almost so by judicious control of the refractive index of the material from which it is constructed. This is admittedly a speculative line of thought so far as we are concerned at the present time. For a highly advanced race, however, it might not be an impossibility.

Before commencing this book the writer scrutinized at length a considerable number of reports of sightings. Clearly it is relevant to quote certain instances in these pages. The problem is which to choose and to this there is no easy answer. For a start it is essential to eliminate the completely ludicrous ones—personal rides in flying saucers and the like. We can dispense also with those obviously arising from misinterpretations of natural phenomena. Undesirable too are those which, though dramatic, cannot be substantiated by other witnesses. This is not impugning dishonesty in such cases—merely recognizing the fact that fear, panic, and hysteria can produce peculiar and often remarkable results.

There is one sighting which certainly must be mentioned. Whatever the merit or demerits of this case it was without much doubt the forerunner of all the others. For that reason alone it has a unique place in "UFOlogy." The precise date was June 24, 1947. It was the day on which Kenneth Arnold, an Idaho businessman, was on a routine flight in a private aircraft to Washington. It was also the day on which Arnold was destined to make history. As he flew over a snow-capped range of mountains he was astounded to observe a number of very peculiar-looking aerial craft apparently flitting in and out between the mountain peaks. The "things," craft, apparitions, or whatever they were made no attempt to interfere with or molest Arnold's airplane. It was probably Arnold himself who unwittingly bestowed on the objects their now familiar title, for in his report on the incident he described them as "saucer-like things." The news of the event soon spread and in no time at all banner headlines throughout the United States and in several other countries were posing the question "Is Another World Watching Us?"

Just precisely what Kenneth Arnold saw above Mount Rainier that June day three decades ago may never be known, but assuredly he saw something. Were the objects really the vanguard of a reconnaisance fleet from another planet which has been watching us ever since or were they merely the result of an optical illusion? It is possible when flying over mountainous regions to witness the sort of optical effect in which a number of circular disc-like apparitions appear apparently just above the peaks. It is purely the result of refraction (the bending of light rays). It is not altogether uncommon, although conditions have to be just right, and indeed the writer has seen the phenomenon on one or two occasions. Was this the type of thing witnessed by Kenneth Arnold? This might be the acceptable explanation but for the fact that such "objects" are most unlikely to flit about in the manner he described.

We must now jump to January 7, 1948, a day equally memorable in UFO lore. On that day there appeared over Godman Field, Fort Knox, a mysterious object which, real or unreal, terrestrial or alien, was destined to cost a young flier his life.

The day was sunny, clear, and cold with routine flying training in progress. Normality, however, was soon to end for something lay above the airbase—and it was huge! From the ground it was easily seen with unaided eyes. "A huge ice-cream topped with glowing red" was one graphic description of the "thing."

Four P-51 pursuit ships of the National Guard were at once ordered into the air. From that of Capt. Thomas Mantell came the strident words "There it is!" None of his fellow pilots had apparently yet seen it. By now the "thing" was climbing rapidly. "Following up to 20,000 feet," radioed Mantell. "If I'm not any closer by then I'll abandon chase."

These, so far as is known, were the last words ever spoken by Captain Mantell. Calls to him remained ominously unanswered and shortly afterward the wreckage of his machine was found, his body still in the cockpit. Meanwhile the cause of all the furor, still not located by the other three fliers, had miraculously faded away.

Two vital questions at once arise: What happened to Mantell and just what *did* he see? Two possibilities have been suggested.

a. He got too close, saw too much, and was ruthlessly destroyed.
b. The object he saw was merely a "sky-hook" balloon or a "mock sun" caused by ice crystals in the atmosphere above him. Then, due to the excitement of the moment, he omitted to turn on his oxygen supply, blacked out, and crashed before he came to.

There are a number of points worthy of comment here. If the object really was a balloon, then its rate of climb can only be described as phenomenal. Moreover, its reported dimensions and metallic appearance are hardly what we would expect. And if the thing was really as large as is claimed and so prominent, then why did the other three pilots fail to locate it? From all accounts it was seen by many on the base itself. Was its rate of climb really so rapid or did it, for reasons we cannot imagine, simply disappear after Mantell's fatal near-approach?

The more one ponders these questions the more arise. Captain Mantell was a qualified, competent, and experienced officer, trained to observe accurately and react logically. He cannot have been unfamiliar with "sky-hook" balloons. The same applies in the case of a "mock sun," and anyway a "mock sun" would hardly have been cone-shaped and of a "metallic appearance." And is it reasonable to expect an airman of Mantell's experience and training to commit the elementary error of forgetting to switch on his oxygen supply?

The official record states simply that Captain Mantell met his death chasing a balloon. Presumably an official record has to give some reason but in the circumstances such a dry, laconic verdict seems both weak and ludicrous.

Five years later, during November 1953, two Royal Air Force officers in a British Vampire two-seater interceptor at 20,000 feet over southern England established contact with something of a very similar nature. This too shot up out of reach and was soon lost to sight. The British air ministry, like its American counterpart, also attributed the phenomenon to a weather balloon. It did not, however, explain how such an object could climb so rapidly as to outdistance a fast jet pursuit aircraft.

Let us now jump to 1956 and this time to a report by Soviet airmen. This in itself constitutes an odd twist. So far, while UFO reports were coming in thick and fast from all over the world, the USSR had said nothing. In more than a few Western minds this led to dark suspicions. Were the Soviets up to something? Were they in some way responsible for these mysterious craft? The report of this particular sighting, like most subsequent Russian ones, was deliberately delayed. The facts, according to official Soviet records, are as follows.

During a polar flight, navigator V. I. Akkuratov and his fellow crewmen, while overflying Greenland, noticed a large object rapidly approaching their aircraft on the port side. He likened the object to a lens, adding that it had "edges which appeared to pulsate." The pilot of the aircraft was compelled to take evasive action in order to avoid collision. This drove the aircraft into a thick bank of cloud. On emerging from this about half an hour later the crew found the object still there, almost as though it had been awaiting them. At once it began to fly alongside them on a parallel course. In an attempt to frustrate the thing the pilot of the Russian aircraft put his machine through a series of maneuvers. These were to no avail for the mysterious object merely matched these to remain on station on their port side. Twenty minutes elapsed during which this tantalizing aerial impasse continued. Then suddenly and without warning the peculiar object accelerated violently and was lost to sight within seconds. All crewmen in the Soviet plane were emphatic that it had no distinguishing surface features. Neither did it reveal the presence of jets nor did it leave any sort of vapor trail. All agreed that the speed of the thing's departure was phenomenal.

It might reasonably be suggested that the incident was entirely due to some optical phenomenon brought about by prevailing weather conditions and, as a consequence, the apparition followed the aircraft just as the image in a mirror follows exactly the movements of the person or object responsible for it. The idea is immediately invalidated by reason of the thing's rapid initial approach and by its equally rapid departure.

As we have already tried to convey, this book is not intended as yet one more about "flying saucers." Therefore it is not the intention of the writer to bore readers with an endless rehash of sightings, supposed sightings, and imagined sightings. Those already mentioned are felt to be of special interest either because they are the first or because they are reasonably well authenticated.

It would seem that the chances of a spacecraft from another planet entering our atmosphere, hovering at low altitudes, or landing without being detected must be very slim indeed considering the amount of electronic surveillance now maintained over our world. Modern radar is fantastically sensitive and likely to become even more so.

There is another highly relevant point. UFOs are at times credited with supersonic speeds. Yet there are no known records of sonic booms emanating from them. On the contrary, these objects flit across the our skies like phantoms, often as not in complete eerie silence. And it goes without saying that their incredible gyrations, maneuvers, and bursts of acceleration would rapidly bring about the total disintegration of any normal aircraft.

Neither of these points applies, of course, if UFOs *are* the product of a society with a technology millennia ahead of our own. In such circumstances it is quite invalid to use criteria based on current terrestrial standards.

Let us recross the Atlantic once more to England for a well-substantiated and inexplicable *series* of events. In Wiltshire, one of the southern counties, some very odd things have been happening over the past few years. Wiltshire is the county which contains the ancient circular stone monument known as Stonehenge, popularly attributed to the Druids, though there are doubts about this. Certainly it is very old and precedes the period of recorded history. Not far from Stonehenge lies the pleasant town of Warminster, a town which for some unaccountable reason has apparently a strong appeal to UFOs. So noted has it become in this respect that it is now something of a mecca for UFO enthusiasts the world over.

For some time after 1964 the townspeople were subjected to odd crackling sounds reminiscent of electrical discharges, thunderous rumblings above the roofs of their houses, the appearance of huge torpedo-shaped objects in the sky overhead, and at times also brilliant lights and balls of incandescence. The whole affair is too extensive, too prolonged, to be the work of practical jokers no matter how dedicated or skillful. Neither can it be attributed to imagination or hysteria, for those who tell of these things are honest, reliable citizens having no reasons whatsoever to concoct silly stories. Whatever the answer, something very odd indeed has been affecting Warminster. But why Warminster? The place is a perfectly normal English county town amid lovely surroundings. It contains no vital secret research establishments, no atomic centers, no rocket-proving ranges. It is certainly a very old town, dating back to Roman times. Later it was involved in the battles between the English King, Alfred, and the early Danish invaders of Britain. It is mentioned in the famous Domesday Book as Guermestre. Later, around the year A.D. 900 it is referred to again, this time as Worgemynster.

The accounts are many and some are quite remarkable. Take, for instance, the case of Kathleen Penton. One evening as she was sitting by an upstairs window, an amazing apparition crossed the sky in front of her. She declares firmly that "portholes" ran along the whole length of the thing, which was of "enormous proportions." According to Mrs. Penton, the "portholes" were illuminated by an intense yellow light. Each end of it was rounded and she likened it to a railroad passenger car. Also, it seemed to be moving obliquely.

What can one make of this? An aircraft at a very low altitude? If so there should have been engine noise. But there was none! An airship, perhaps, drifting with its engines idle? Only there are no such craft around today—and have not been since the ill-fated German zeppelin *Hindenburg* crashed and burned at Lakehurst, New Jersey, on a thundery May evening back in 1937. In this day and age most people are reasonably familiar with the contours and general appearance of aircraft. If this were a conventional aircraft, where were its wings, tail planes, and fin?

In June 1965 the wife of a local minister described how she had watched a brilliant, glowing, cigar-shaped object in the sky for fully twenty minutes. In this instance the whole family was able to see the spectacle. Use of a small telescope which was handy showed that the thing stood vertically in the sky, the lower end being thicker than the upper. The lower end also appeared to have some sort of ring arrangement girdling it. The entire object appeared to be in slow rotation around its vertical axis. By September of the same year upward of two hundred residents claimed to have seen this peculiar object at one time or another. According to the records, the appropriate authorities insinuated that it was merely "a flock of geese." One has the feeling that the only geese around were the people who could come up with such a fatuous explanation!

It has been suggested that Warminster or its environs may at some time in the past have been the unsuspecting recipient of a homing device planted by the occupants of a UFO so that future extraterrestrial wayfarers might thereby get their bearings. This is not a very convincing explanation but at least it makes more sense than the aforementioned flock of geese!

The result of all these goings-on and the subsequent publicity has been to make Warminster a magnet for most of the UFO addicts in England–and from all accounts a goodly number from the United States as well. Members of the press and of even more august bodies have turned up to see what is what. Not all have been disappointed, which renders the Warminster mystery even more remarkable. Normally in instances like this outside observers expect nothing–and see precisely that!

There is on record a most interesting account of a visit to Warminster by Robert Chapman, Science Correspondent of the London *Sunday Express.* On being informed that a good vantage point for observation of the town and its environs was a hill known as Cradle Hill, he proceeded there and climbed to the top. It was a May afternoon but from all accounts a most thoroughly un-English one with a fine drizzly rain falling steadily. Only two things happened: He heard a cuckoo and got soaking wet. Neither of these could be attributed to aliens! However his spirits were not dampened, even though his clothes were, and in the evening he returned to the summit of Cradle Hill in company with a local resident. By then the rain had ceased, though the sky remained overcast. The time was just after 11 P.M. Darkness had fallen and a fairly strong breeze was now blowing. As they waited atop Cradle Hill the sky cleared in patches and occasionally a star or two could be seen. Once a jet passed across the sky above the cloud banks but, apart from this, there was no sound save the sighing of the wind. By midnight both Chapman and his companion began to feel restive. Clearly they were on a useless errand and both men began to feel there was little to be gained by a continuation of their vigil.

They had just begun to move toward the road leading down the hill when suddenly they were confronted by a peculiar, amber, pulsating light in the sky. Chapman is emphatic that it was too high to have any possible connection with road traffic, and certainly it was much too large and bright to be a star. The total absence of any aircraft sound ruled out that possibility also, apart from which the mysterious light simply hung motionless in the sky. It disappeared after a few minutes but soon reappeared in exactly the same spot. Then it vanished again, only to reappear accompanied by a pulsating companion. Chapman considered the possibility that the phenomenon might just be due to car headlights on a vehicle parked, perhaps, on an incline some distance away. On reflection he came to the conclusion that this could safely be ruled out since they were much too bright, pulsated, and were too far apart. Both lights then disappeared, but shortly after a single pulsating light appeared briefly once again. After this had gone both men hung around the summit of Cradle Hill for a little longer, but seeing nothing further proceeded back down the hill toward the warmth, light, and seeming security of Warminster.

So much then for Warminster and the strange mystery and aura which has (all too literally perhaps) descended upon it.

The facts seem to be well corroborated. And the verdict—? Either an extreme case of mass hysteria, or *something is there*! There can be little doubt that what has been happening in the skies over this peaceful little Wiltshire town over the past few years is very odd indeed and there is no indication as yet that this mysterious affair is at an end.

Another mysterious event which took place in October 1967, also in England, is worthy of mention. This is another occasion where there is certainly no dearth of witnesses. The actual locality was near the city of Exeter in Devon, one of the southwestern counties of England.

The story begins early on a very cold morning. The first

witnesses were two members of a police patrol who suddenly saw the "thing" from their moving car. Indeed, they could hardly miss it for it was glowing, though not brilliantly. The source of this odd radiation was apparently sailing along above the treetops ahead of their car and had neither the appearance nor the motion of any conventional aircraft. As the patrolmen increased speed in an attempt to draw closer, the object was disobliging enough to do precisely the same thing. This escalation of the chase continued until the police car was doing 90 miles per hour with the unidentified object still well ahead. Finally as the police car dropped back the object was seen to be joined by a second, the latter being large, extremely brilliant, and shaped like a cross. Air Force stations in the surrounding area denied all knowledge of the thing and stated quite flatly that none of their aircraft were then airborne—and that anyhow they had none that looked like that—and hoped they never would!

If at first the two patrolmen were feeling puzzled and foolish, vindication of their claims was not long in coming. Within the next few days more and more reports came in of peculiar "craft" and other manifestations in the skies. Indeed, there was quite a wide selection from which to choose. This included cigar-shaped objects, disc-like objects, and even more significant in the circumstances, huge and brilliant "flying crosses." It seems reasonably safe in this instance to rule out publicity seekers and the lunatic fringe. A "flying cross" (it makes a change from a "flying saucer") was witnessed by no fewer than six police officers in Derbyshire, a county about 200 miles to the northeast of Devon. Farther north still, in Scotland, coastguards in Wigtownshire (only about 150 miles from the writer's home) reported a "great silver cartwheel" which appeared over the coast from the west doing an estimated 400 miles per hour and emitting a peculiar electrical whine. The thing was in full view for about ten minutes before disappearing over the northeast horizon. Around the same time an Air Force wing commander and his wife driving in Hampshire (south coast of England) were astounded to see seven bright lights in "V" formation sailing majestically and silently across the sky. Even as they watched, the objects changed formation and assumed the configuration of the increasingly familiar cross. The color was pale yellow and seemed to pulsate. (A flock of geese on night patrol?)

Here, then, are instances of report by police, coastguards, and a senior Air Force officer—hardly the types of persons to be easily misled or likely to concoct crazy yarns about UFOs.

The authorities at the Royal Observatory, Greenwich,* were clearly puzzled. Feeling no doubt that they had better say something, they issued a statement that indicated the planet Venus as the cause of all the turmoil. Not that they were trying to suggest that what had been seen were spacecraft from that planet—merely that Venus itself, hanging in the skies of Earth some 30 million miles distant, was the object responsible. As an explanation it ranks on a par with that flock of geese! When it was pointed out to them that the objects had been seen *under* a cloud base they agreed that Venus could hardly have been responsible and proceeded to deliberate further. The conclusion they reached cannot be termed momentous. "There is something up there," they stated unemotionally, "that is neither star nor planet!" Many had already come to that conclusion.

The fact of reliable witnesses was invoked recently by none other than U.S. Senator Barry Goldwater of Arizona. Senator Goldwater, himself a retired Air Force Reserve brigadier general and a pilot of almost fifty years' experience, said, "I've doubted stories from many witnesses, but when qualified airplane pilots and other experts tell me they've seen strange, unexplained flying objects I have to put faith in their reports. Air Force, Navy and commercial pilots have revealed to me cases when a U.F.O. would fly near them—right off their plane's wing—and then just zoom away at incredible speeds."[2]

In a number of instances UFOs have been seen to land—or at least some inexplicable apparition appears to have touched down. As often as not there is no visible aftermath of the object's descent. At this point we might consider the converse—instances in which peculiar impressions *are* left on the ground though *no* UFO has been observed.

An interesting example of this sort of thing is the peculiar damage done to a couple of barley fields on the Isle of Wight one

*This observatory used to be at Greenwich, at which time it was known as The Royal Observatory, Greenwich. Due to light and smog it was moved to Sussex, where it is now known as "The Royal Greenwich Observatory."

night in 1967. The Isle of Wight is a roughly triangular island just off the coast of southern England very close to the city and great seaport of Southampton. The fields concerned are separated from an adjoining main road by a hedge. Parallel to this hedge there appeared overnight a most remarkable series of tracks. "Tracks" is not perhaps the most appropriate term for the peculiar markings. They were in fact a series of closely linked roughly circular depressions in the growing barley. The owners of the field were completely mystified. They had neither seen nor heard a thing and declared categorically that there were neither people, animals, nor agricultural implements in the vicinity of the fields at the time. Neither had there been any storms of late. For days thereafter both agricultural and aeronautical investigators examined the damage, which extended for approximately three-quarters of a mile. One of the latter declared that in his opinion these kinds of marks could have been caused only by some flying object attempting to land.

Emphasis was added to this belief by a report from a number of schoolboys. However, since this came a few hours *after* the event became known, it has to be treated with caution. The boys claimed that on July 10, 1967 (just prior to the business of the barley fields) they had seen "a white milky disc hovering in the sky." They claim to have seen it again, two hours later, falling this time "like a leaf" and apparently not under complete control. After a minute or two it seemed to steady but continued to descend. At this point, due to the contours of the ground, it passed from their view, but in the general vicinity of the barley fields.

The official report on the incident by the team of aeronautical investigators spoke of areas among the barley which had an almost circular shape. The orientation of the barley was clockwise in some of these areas, anti-clockwise in others. Some of the barley was broken and there were even parts in which the roots had been gouged out.

Straying cattle, rampaging dogs, and freak winds were easily ruled out. These things could and often did damage crops but always in the most haphazard way—never in this remarkable manner. A practical joke perhaps? A pointlessly involved one if it was and not exactly easy to carry out.

Some years earlier, in July 1963, there had been a similar episode. On this occasion it was a potato field which was involved.

The nature of the damage was a bit different, however—a crater 8 feet in diameter. Once again the county was Wiltshire. There was absolutely nothing to indicate what had caused the crater. It had simply appeared. Now craters do not just appear. Either something scoops, blasts, or gouges out the soil or rock, or a subsidence causes a cave-in. The field's owner had apparently no doubts as to the identity of the perpetrator. "It could only have been caused by a spacecraft," he declared emphatically. Not surprisingly his views engendered little support among his immediate circle!

An army team was subsequently moved in to investigate. They started to search around but nothing of significance showed up—not, that is, until they started to employ detecting instruments. These clearly indicated the presence of metal. At this point the mystery began to escalate for two other "craters" of a similar nature were now found in adjoining fields. And just for good measure some apparently "burned" vegetation was discovered in a hedge adjoining one of the fields. Had this been a case of a "bouncing" flying saucer?

The army now commenced to dig. A meteorite seemed a logical explanation for the mystery though none had been seen or heard to fall. Now meteorites are not exactly shy about announcing their arrival. Quite the reverse in fact—such descents are generally spectacular with respect to the light and sound effects they produce.

All that was found in the "crater" was a chunk of rock weighing about half a pound, which on a first, cursory examination looked not unlike a stony meteorite. Subsequent geological scrutiny, however, proved it to be a chunk of local rock which had every right to be there. And a further disappointment to UFO aficionados, biological examination indicated that the supposedly "burned" vegetation was vegetation that had probably been attacked by mildew. The Army departed, satisfied that its time had been well and truly wasted. The official report stated that no conclusive evidence could be found for the existence and sudden appearance of the crater. It was not due to a bomb, exploded or unexploded, and it was *not* due to a meteorite. Neither could its origins be attributed to subsidence.

It now came out that on July 10, a fortnight before the discovery of the mysterious crater, a member of the local police force had reported what he took to be a very bright meteor flash across the sky and disappear suspiciously near the field in which the crater had appeared. If this *had* been a meteor fall, why were there no remnants? And if it were not a meteor—?

Peculiar happenings in the area surrounding the lovely island of Bermuda and off the east coast of the Florida peninsula have recently been the subject of much interest and speculation. These incidents relate more to *dis*appearances than appearances and concern both ships and aircraft. It might be as well to state, however, that though there is undoubtedly an element of mystery in that region, which has come to be known as the Bermuda Triangle, some writers have endeavored to generate even more by a slanting of events.

A classic instance is that of the now almost legendary flight of Grumman Avengers on a training mission, which disappeared without trace—as did a large Martin Mariner rescue plane sent to look for them. It would seem, however, on the available evidence that the planes of Flight 19 simply became confused and, seeing certain islands off the Florida coast to the east believed these to be the celebrated Florida Keys. Thinking therefore, since they were on a westerly course, that they were overshooting the Florida peninsula and heading toward the Gulf of Mexico, they turned through 180 degrees and began flying on a reciprocal course. This actually sent them out over the Atlantic, where eventually their fuel ran out. The case of the missing search plane is now explained by an explosion. This machine, as befitted its purpose, carried a large load of gasoline and fumes from this were known at times to filter into the plane. A spark could easily have done the rest. A brilliant flash, supposedly emanating from the last reported position of the plane, has been reported. The oddest feature of the entire incident is the fact that no wreckage from any of the aircraft was ever found despite an intensive sea and air search of the entire area.

There have admittedly been reports of odd messages from Flight 19. A comment by one of the pilots is said to have been "Don't come after me . . . they look like they are from outer space!" Another is said to have said that everything about the sea "looked wrong." If these reports are true (and it does seem rather a big "if") then the whole incident takes on an entirely new dimension—and a rather sinister one.

Even less easy to explain in conventional terms is the case of the Tudor airliner *Star Tiger* of British South American Airways which vanished in perfect weather only 380 miles from touchdown in Bermuda. No distress call was ever received from the aircraft nor was any vestige of oil or wreckage ever sighted. The machine to all intents and purposes was simply swept out of existence. This is reflected in the cold, formal, legalistic jargon of the official court of inquiry:

> It may be truly said that no more baffling problem has ever been presented for investigation. In the complete absence of any reliable evidence as to either the nature or the cause of the disaster to "Star Tiger" the court has not been able to do more than suggest possibilities, none of which reaches even the level of probability. Into all activities which involve the co-operation of man and machine two elements enter of very diverse character. There is the incalculable element of the human equation dependent upon imperfectly known factors and there is the mechanical element subject to quite different laws. A breakdown may occur in either separately or in both in conjunction. *Or some external cause may overwhelm both man and machine.* What happened in this case will never be known!

Within a year a sister aircraft, *Star Ariel,* identical in all respects, disappeared in similar circumstances in the Bermuda region. Once again there was no distress call nor any trace of oil or wreckage.

These aircraft are only two of several which have inexplicably vanished in this region. Ships have also disappeared unaccountably—all kinds from light pleasure craft to large ocean-going vessels. And always the common factor is there—no distress call, no wreckage.

One of the most intriguing episodes to occur in this region involved neither UFO appearance nor conventional aircraft *dis*appearance. Earlier we examined time warps and time annihilation. Consider, then, the strange case of a National Airlines Boeing 727, in 1970, inward bound for Miami, Florida, from the northeast. This is one of these instances which appears to be well documented and verified. The aircraft was making a normal approach when suddenly and unaccountably it disappeared from the radar screens of Miami air traffic control. Inevitably this caused alarm. Ten minutes later, however, the blip reappeared on the screen and the aircraft touched down safely soon after. Pilot and crew were surprised at the fuss since, so far as they were concerned, all had been perfectly normal. It was then that a very curious anomaly came to light—clocks on the aircraft and watches belonging to the flight-deck crew were all uniformly ten minutes *slow.* A further check revealed that the watches on the wrists of all the stewardesses showed exactly the same discrepancy. Discreet additional inquiries among the passengers showed the same thing.

This could not possibly have been due to coincidence. How, then, to account for such a curious state of affairs? It must be assumed that some physical force or agency had affected all the clocks and watches aboard the plane. And for some physical reason the plane had been removed from radar surveillance for the same ten minutes. Had some alien agency created a vortex or "tunnel" in the space-time continuum and having done so decided after all to return the machine? Certainly passengers and crew noted nothing amiss—or had memories of that missing ten-minute period been mysteriously expunged from their minds?

As mentioned earlier in the chapter the writer had occasion to fly through the region en route from England to Miami in July 1971. At that time the so-called Bermuda Triangle had not been widely publicized. Certainly the writer was blissfully unaware of its peculiar and sinister reputation—which in the circumstances was probably as well. As it turned out, the flight was quite uneventful save for one trifling feature—trifling, that is, at the time. "Intriguing" would now be a more appropriate word.

It was while watching the thunderclouds rise from the sea that the writer saw another aircraft become visible at a much lower altitude. What, the reader may be tempted to inquire, is so unique about that? Nothing, save for the fact that this particular airplane was about fifty years out of date. It was, in fact, an absurdly old and antiquated aircraft to be flying over the ocean in the summer of 1971—a large silvery biplane with two motors and a fixed undercarriage of four wheels, two pairs of two abreast (seen when the machine banked steeply). In all respects it closely resembled an airliner or military machine of the early 1920s. It flew into one of the clouds—but not apparently out of it!

There may be nothing significant in this whatsoever. Perhaps some group of aviation enthusiasts along the Florida coast possess a serviceable aircraft dating from that period. Nevertheless, the memory of the incident continues to intrigue. The plane was such an anachronism and during the few minutes it was clearly visible there certainly seemed to be a peculiar, unsubstantial, tenuous look about it. This, of course, could simply have been due to haze. It appeared almost as if it had been "switched on" and, as mentioned, was not seen to emerge from the other side of the small puff of cloud which it entered.

Now, with the benefit of hindsight and knowledge of the alleged peculiar propensities of the region, it is very easy (and tempting) to let the mind wander to the subject of space warps and the like, to wonder if a machine of that type did half a century or so before disappear there, to fly forever as a sort of aerial Flying Dutchman, half here and half "there"—the helpless plaything of alien minds. If at some time in the future reports from the area speak of a mysterious flight of Grumman Avengers which appear from time to time, then the event described would probably come to be regarded as highly relevant!

It is interesting also to reflect on a very peculiar incident reputed to have occurred in Biscayne Bay, Florida, one evening in September 1973. A Miami college lecturer and three companions were returning from a fishing trip in the bay. Darkness was just beginning to fall. Suddenly they realized that the compass in the launch was about 90 degrees out. Almost at once the lights on the launch dimmed and soon went out altogether. However, the shore lights were in full view so that the lack of compass and even of the lights was merely a nuisance. Their course was due west. It soon became apparent, however, that even with the engine at full power no progress was being made—not, that is, to the west. For some odd reason the craft was gradually being driven to the north.

It was then that the four occupants noticed, to their considerable disquiet, a *large dark shape* obscuring the stars to the west. As they watched this and discussed its probable nature they clearly saw a *moving light* approach it then disappear. Within a short time the dark mass also disappeared. Then and then only did the compass revert to normal and the launch begin to make normal progress to the west. Around the same time others in the area reported similar experiences.

Now had the compass not gone awry and the launch not been

pulled mysteriously northward, the effect in the sky might simply have been attributed to a cloud and the moving light to an aircraft passing behind it. It could be claimed, of course, that the northward drift was an optical illusion. It is difficult, however, to explain the 90-degree gyration of the compass and the fading and eventual failure of the lights. It should be added that the engine of the launch was a diesel and therefore remained unaffected by failing electrical equipment. Equally difficult to explain is the sudden recovery of both compass and generators after the dark mass in the sky disappeared.

A rational explanation for the disappearances and other queer manifestations in the Bermuda Triangle is difficult to come by. The word "coincidence" springs to mind but just as quickly fades. There have been too many such events for it to be valid.

If these occurrences should be due to alien activity, there still has to be a physical explanation of some sort. Aliens may be highly advanced technologists but they are unlikely to be miracle workers.

It has been suggested that UFOs somehow create a strong temporary vortex and that it is such vortices that have caused ships and aircraft so mysteriously to vanish. This invokes the "unified field theory" of Albert Einstein. Most of us tend to regard the entities of space/time and matter/energy as entirely disparate things. Now there was a day when space and time were themselves seen as unrelated entities. Today we accept the relationship between them. This was equally true of matter and energy until the well-known equation $E = mc^2$ put matters into their correct context. In a unified field concept space/time and matter/energy themselves become interchangeable—or more correctly transmutable. Such a concept might explain not only the disappearance of ships and aircraft but also the peculiar way in which UFOs themselves allegedly appear and disappear. The fundamental force, as might be expected, is electromagnetic. A current flowing in a coil of conducting material produces both an electric and a magnetic field, the one being at right angles to the other. This invokes the existence of two planes in space. We realize, of course, that there are three. Is the third—represented by a gravitational field produced at will by the application of a massive electromagnetic effect—the "vortex" linking the entities of space/time and matter/energy, thereby rendering possible the ready transfer of matter through space? There have been from time to time unsubstantiated reports of experiments along these lines by scientists on Earth.

Inevitably such thoughts tend to tie in with the idea of breaking through the normal "skin" of space-time into another dimension—the "shortcut" of "non-space" outlined in Chapter 3. We are not claiming that this is actually happening in the Bermuda Triangle. Nevertheless, if all these peculiar events have a common origin and that origin is due to the activities of alien beings, then we have at least something which cannot be entirely disregarded.

The next question is why? Why should aliens commit such acts? The answer could be brutally simple. We are being sampled! We dredge samples of marine life from the ocean depths in the interests of knowledge and scientific research. There seems nothing ethically wrong about it—not at least to us! If the creatures dredged up were capable of coherent thought they probably would have rather different ideas on the subject. A highly advanced alien culture from the stars might regard us much as we regard these lowly marine creatures. This is not a matter of right or wrong, of ethics—merely one of position in the great cosmic pyramid of life!

A close look at the subject of UFO sightings does seem to indicate that the number of reports emanating from the general Florida Bermuda region is very considerable. Indeed, it would appear to be in excess of that for any other specific region. Coincidence? Perhaps—and then again perhaps not. Who can say? Statistics can be misleading, yet they can never safely be disregarded.

The proximity of the launching pads of Cape Canaveral and the Kennedy Space Center has been cited as a possible reason for excessive UFO activity in this region. At a first glance this seems a reasonable sort of supposition. There are, however, a couple of points which effectively downgrade this idea. If there is alien activity in the region, it represents a capability in space travel about as far removed from that going on at the Cape as an intercontinental ballistic missile is from a bow and arrow. Surely an advanced alien technology would find little of interest in our first tentative probings of the space around our own planet. We must also remember that the peculiar events occurring in the area were taking place long before the first rockets lifted from the eastern shores of Florida.

UFO sightings seem to come in waves. The halcyon years were undoubtedly those immediately following World War II. There was another rash in 1957 and again in 1965-1966. Whether or not there is anything significant about these dates it is impossible to say. It would now seem as if 1973-1976 must be added to the list. In the United States alone reports have come in from places as widely separated as California, Mississippi, and Vermont.

For one of the most spectacular of this recent series we must go to Chattanooga, Tennessee, on the night of October 17, 1973.[3] The report tells of a "hissing, glowing, cigar-shaped U.F.O." which hovered in front of two astounded policemen and a group of about twenty-five persons before zooming quickly out of sight. Now one person might be fooled, even two or three, but hardly twenty-five. One of the policemen, himself an experienced private airplane pilot, said, "An object making a hissing sound like that of escaping gas was hovering twenty feet above a wooded swamp. It was about thirty to forty feet long and ten feet high. It was below tree-top level and was clearly framed between two trees. A brilliant blue-white glow surrounded the whole thing. I have never before seen a light of such intensity. Certainly this was no airplane."

His fellow officer substantiated the description, adding, "It was like a cigar lit at both ends." Five miles away another patrolman saw it too. "I spotted a brilliant white light low in the sky about 8:30 P.M. It lit up the whole neighborhood and I began to wonder if my mind was playing tricks on me. I tried to follow it but it vanished. It certainly wasn't a plane, a helicopter, a balloon, a star or the Moon."

Assuming all the facts are as the witnesses stated, then there is simply no rational explanation for a phenomenon of this sort other than assuming extraterrestrial powers at work. Methane ("marsh gas") often emanates from swamps due to decomposition of vegetable matter. This has been known to ignite spontaneously, giving rise to the effect known as will-o'-the-wisp, but never to produce an effect like that just described.

There have been numerous reports of peculiar creatures either emerging from UFOs or being seen in the vicinity when one has reputedly landed. Most of these are highly suspect at best and for that reason the writer has preferred to ignore such instances. However, an exception should perhaps be made with respect to one of these for reasons which will be stated later.

This particular incident took place in Mississippi in the vicinity of the Pascagoula River on October 11, 1973.[4] It involved two men who had just completed a pleasant afternoon's fishing. It had just begun to grow dusk when both noticed a strange blue light floating toward them across the water. As it grew closer they perceived that it was an "aerial craft" of some kind. At this point a hatch is said to have opened in the side of the thing. At this open hatch both men claimed they could see "three wrinkled creatures with pointed ears, lipless mouths and crablike hands." One of the men immediately fainted (who can blame him?). The other, apparently made of sterner stuff, held his ground until the craft drew alongside. Thereupon one of the creatures is said to have touched him, thereby rendering him weightless. The aliens then took him on a conducted tour of the strange vessel.

Now this story sounds like one of the worst from the 1947 crop or one of Hollywood's less successful efforts. Normally and rightly it would be dismissed as so much sensational rubbish. Only there is a "but" here—rather a vital one.

The two men were later interrogated by no less an authority than Dr. J. Allan Hynek, professor of astronomy at Northwestern University.[5] The interrogation was carried out under hypnosis—a highly significant point. As a result of this Dr. Hynek is adamant that the two men were indeed telling the truth. Commented Dr. Hynek, "It was like an aborigine who sees his first Boeing 747. U.F.O.'s are real phenomena that are just as puzzling to us as nuclear energy would have been to Benjamin Franklin."

After this sighting, reports of others started to escalate. Another noteworthy example is the experience of Ohio State Governor John Gilligan and his wife, who watched for half an hour a vertical amber object unlike any known bird or plane.[6] One would hardly expect a state governor to make up a story like this, so clearly Mr. and Mrs. Gilligan either saw something very peculiar or misinterpreted some other aerial phenomenon. From Marin County in California [7] two sheriff's deputies (among several others) reported "a brilliant light with an orange tail streaking across the sky." This might just have been a meteor, although the description is a shade too spectacular for that.

About the same time a very frightened family told of an object like a "lit-up house" that landed near their home in Gulfport, Mississippi. Presumably this one took off again. At least there is no record of any occupants having emerged. Near Cleveland, Ohio, an experienced helicopter pilot had to take violent evasive action when "a cigar-shaped craft with a red light" came directly at his machine. Said the badly shaken pilot afterward, "We never saw anything like it. It was unreal!" [8] It is noteworthy how often the description "cigar-shaped" comes up.

Now if all these people are telling the truth (and for the most part they do not seem the kind likely to fabricate absurd tales) then the situation can only be described as intriguing—perhaps even disturbing. It is virtually impossible to dismiss reports of this nature as being due to misidentified aircraft, weather balloons, bright planets, or stars.

On the other hand, as noted astronomer Carl Sagan has said, if all these reports are due to extraterrestrial craft from the stars it means that a host of alien civilizations must "have us in their sights." In view of the inexorable parameters of time and distance this at first glance does seem unlikely. But, as we saw earlier, the highly sophisticated techniques of advanced galactic communities could have drastically downgraded such parameters. Just one civilization with such powers could be responsible. There remains, too, the possibility of gradually encroaching colonization by aliens which we mentioned earlier. It is as well to bear in mind also that a terrestrial society which has in the last three decades developed nuclear weapons and a primitive space capability could be regarded as one to be watched—and watched closely!

So much, then, for the contemporary scene. What further can one say? What verdict is possible? The writer, for his part, must confess to a time when quite arbitrarily he dismissed all UFO reports as simply due to some peculiar hysteria of the times. Now, very sincerely, he believes this to be no longer possible. Undoubtedly many reports are fakes, undoubtedly many are false, but all do not come into these categories. Something decidedly odd may be happening out there and in the lifetime of many of us the answer could be revealed.

It is of interest once more to quote Senator Barry Goldwater, who in a recent interview summed the situation up rather well. "Some day soon," he said, "someone's going to have strong U.F.O. evidence that can't be explained away. I believe in U.F.O.'s and if my secretary came in one day and said, 'Senator Goldwater, there's a three-foot tall green man with four eyes and an antenna sticking out of his head waiting to see you outside,' I wouldn't think the poor girl was nuts. I'd simply say to her, 'Send him in!' "[9]

Reports indicate that 1976 saw a further resurgence in U.F.O. activity. These reports emanate from many parts of the world and from a wide variety of people.

Probably the most notable instance of the period was an alleged sighting by the future President of the United States, Mr. Jimmy Carter,[10] although now there are reports that this may have been a case of mistaken identity of a celestial object, probably a planet.

Said President Carter, "It was a very peculiar aberration. But about twenty people saw it. It was the darndest thing I've ever seen. It was big, it was very bright, it changed colors and was about the size of the Moon. We watched it for about ten minutes but none of us could figure out what it was. One thing's for sure. I'll never make fun of people who say they've seen unidentified flying objects!"

An interesting report also comes from a British police sergeant in the southeast English county of Kent who while coming off duty, noticed a brilliant white light in the sky over Dungeness.[11] He thought that probably air-sea rescue operations were in progress but a hasty phone call to the appropriate control center indicated this wasn't so. The light then vanished and seconds later a Kent radar station reported unidentified aerial traffic in the area, traffic they could not explain.

From Spain comes a reported sighting from the commanding officer of the Spanish Air Force in the Canary Islands.[12] He was at the time in Sabada, a village in a remote area of eastern Spain when he spotted the object overhead. He said: "It was giving off a brilliant light and was traveling at fantastic speed."

Also from Spain comes one of those reports which in normal circumstances would probably be consigned to the nearest waste-basket. The witness however, was one of Spain's most eminent medical men, Dr. Francisco Padron.[13] His account, which is rather disturbing, is as follows:

"It was nearly dark and I was driving along a lonely road when I saw a round sphere about 13 meters (40 feet) in diameter hovering above the road ahead of me. It was emitting a bluish light and was about 7 meters (20 feet) above the ground. As I approached my radio cut out. I passed right underneath the object and saw, silhouetted inside through a type of porthole, two very tall figures dressed in bright red. Then it accelerated and vanished in the direction of Tenerife."

These are just a few of the 1970s crop. We must ask again if a remorseless tide of alien colonization is steadily approaching this defenseless little planet of ours. It is a disturbing thought and one which continues to intrude.

# 16. FRIEND, FOE—
# OR INDIFFERENT?

We can hardly end our deliberations without giving consideration to the most pertinent aspect of all—the attitude of aliens toward us. In earlier chapters we have only touched on this. Now in closing it is time to dwell upon it more fully. Unfortunately no real answer is possible until the event happens. And by then of course it could be too late!

It is not unpleasant to envisage the possibility of some advanced, humane, and cultured race descending upon our sorely tried and perplexed world and putting it to rights; to contemplate the guiding hand of an "elder brother" from the stars, a mind knowing all the dangers, all the pitfalls—and all the answers! But would it work out that way? Might not the reality prove hideously and tragically different?

It goes without saying that life-forms millennia *behind* us are unlikely either to be signaling or heading our way in sophisticated starships, any more than the dinosaurs or Stone Age men of Earth would have done. But what of races millennia *ahead* of us? To terrestrial scientists life several thousand years behind our own would be a matter for the closest examination. By the same reasoning beings this much ahead descending on our world might regard us also as mere interesting primitives, beings worthy only of clinical laboratory study.

It has been suggested that such a race might just ignore us, regarding us as of little significance. On the whole this seems very improbable. If the purpose of their mission were purely scientific, such a policy would be self-defeating. And were they bent on colonization, they could not ignore us for the most obvious of reasons! If a colonizing expedition from Earth were to descend upon a remote planet today, would it, were the occupants of that world mere happy primitives, treat them with humanity and compassion? Circumstances would no doubt influence this but were the expedition from Earth of the one-way variety due either to sheer distance, a dying Earth, or escape from terrestrial persecution it could all too easily prove a case of "us or them"—the basic law of the jungle. Ethics would in all probability be abandoned at once. Survival, like necessity, is an exceedingly hard master and one that recognizes few bounds. The impact of an advanced civilization on one less so is, as the history books show, generally to the disadvantage and often the distinct detriment of the latter. This may be an unpalatable fact. It is also a very true one.

We must also remember how little there would be in common between societies separated by massive time gaps. An Archimedes, even a Leonardo da Vinci would be utterly out of his depth among contemporary scientists and technologists. Indeed, he would probably find the average layman his superior in terms of knowledge, expertise, and know-how. What, then, of the position between ourselves and a highly advanced society coming to us from the stars? There would be not only this time and knowledge gap of mammoth proportions; there would also be the space gap—a yawning gulf of differing outlook and environment which could so easily prove unbridgeable.

It would seem that the risks entailed in an encounter with a remote alien civilization could be real and very great, the chances of benefit remote, perhaps nonexistent. Indeed, on the subject of electronic communication between galactic communities, Professor Zdenik Kopal puts it rather succinctly. "Should we ever hear the space-phone ringing," he says, "for God's sake let us not answer, but rather make ourselves as inconspicuous as possible to avoid attracting attention!" Inadvertently, though, we may already have called "them." For the better part of a century now radio signals have been emanating from the surface of our planet. Perhaps the first faint crackles and splutters of the crude apparatus of Hertz and Marconi did not get very far, but without a doubt much thereafter was of sufficient power and of the right frequency range to go winging off into the black endless depths of space. Though reduced to the merest of whispers now, these same whispers could quickly and very easily become the loudest of shouts if fed into the ultrasophisticated circuits of alien listening stations. Already some of these signals are over half a century on their way, over 50 light-years out. At this very moment at several points within a concentric "shell" of stellar systems this distance from the sun, alien eyes could be exchanging meaningful glances, alien minds thinking meaningful thoughts—thoughts which, if translated into action, could easily bode ill for this world of ours and its many peoples. Nor is it essential that we restrict ourselves to distances of this order. What of the roving, reconnoitering alien probe or the listening, lurking alien star cruiser? Fantastic? Perhaps—and then perhaps not. A thing like this is fantastic and impossible only until it happens!

Today one frequently hears the term "preemptive strike"—the possibility of attack by one nation on another before the other becomes too powerful. A similar situation could conceivably develop between worlds. Back in the summer of 1939, during the run of the New York World's Fair, a science fiction magazine came out with a cover which, in bright colors, illustrated the consequences of such a possibility. The theme of the Fair was "The World of Tomorrow." Its symbolic centerpieces were the Trylon and Perisphere, the former a tall, triangular shaft tapering to a point 728 feet high, the latter a true sphere 180 feet in diameter. The Trylon and Perisphere represented the structural principles of the column and the dome and were described as "an expression of the shape of things to come." The magazine cover artist had certainly given a new and novel twist to the "shape of things to come"—an alien starship blasting both structures to destruction with lethal rays prior to setting upon the city of New York in earnest. Already the concept of a preemptive strike against developing terrestrial civilization had been seen as a possibility!

Whereas many might not endorse the pessimistic premise of Professor Kopal, few would deny the possibility of some truth in what he says. Much could depend on how remote (or how near) the closest alien civilization happens to lie. This is an estimate on which opinions differ greatly. Figures have been quoted which range from a dozen light-years to several thousand. One of the world's foremost authorities on the subject, Carl Sagan, puts the distance at around the 300-light-year mark, which seems reasonable. If the nearest advanced societies are at this distance, or something around it, then at least two centuries have still to elapse before the expanding "shell" of terrestrial radio waves reaches them. This, however, does not apply with respect to probes or roving star vessels which, their alarm circuits alerted and "ringing," hasten back to their parent worlds by unorthodox, time-destroying techniques.

If danger to us is represented more by an "advancing front" of alien cosmic colonization, then the emission of radio signals from our planet assumes considerably less importance. In such circumstances the aliens, sooner or later, are going to appear in the skies of Earth anyway. But at least they would have been made aware that the inhabitants of Earth were long past the Stone Age and that tactics of a more "persuasive" nature would be required. On a Stone Age planet the natives are mere ants; on an early nuclear one, ants with a decidedly nasty sting! Of course, both types are easily eliminated. It is only a matter of using the correct pesticide!

An obvious and direct threat by alien beings directed against our world and its civilization would, we must suppose, initially at least, confer certain beneficial effects. At the moment we are, have long been, and presumably will continue to be, a world divided. There is nothing quite like a common external threat of massive dimensions to render disputing neighbors suddenly desirous of holding hands and uttering pleasant words (of dubious sincerity) about one another. With the possibility of an alien hell about to cascade from our skies there would be neither point nor sanity in a continuation of the absurd and dangerous confrontations which have so epitomized the past few decades. Even our economic difficulties would speedily evaporate, the most intractable problems suddenly appear petty. About fifty years ago in his celebrated *War against the Moon,* writer André Maurois suggested that a really positive method for securing and preserving peace among the nations and peoples of the world would be to invent a spurious and imminent threat from outer space. Such a scheme, though somewhat impractical, has a certain all-too-obvious merit. It is interesting to reflect on how much more effective would be a genuine one!

There is another possibility—that of opportunism rearing its head, of some group or state developing traces of the "what's-in-it-for-us" syndrome. It is a very short step from that to the "if-you-can't-beat-'em-join-'em" philosophy—the doctrine of ultimate expediency. We might, given the appropriate circumstances, envisage the chance of cunning and clandestine attempts to come to an understanding with the invader in an endeavor to secure favorable terms. In this respect much would no doubt depend on the form, temperament, and aims of the aliens. Were the invaders from the stars too far ahead intellectually and/or too different in physical characteristics any such moves would probably be brushed aside, ignored—even be unnoticed. When applying pesticide who notices the single "friendly" ant?

Thus far we have apparently set our faces against the possibility that invading aliens might be understanding, friendly, and helpful. We have argued that hostility is likely simply because of pressures and the fact that the gulf in terms of space, time, and outlook would simply prove too great. We feel that if a backward society were to stand in *our* way we would remove it because expediency and necessity dictated no other course. What we are really saying is that though we may have come a long way technologically we have made much less progress morally. Looking around our world at the present time this would be a very difficult proposition to dispute. Are we not, however, ignoring the possibility that among some alien races moral and technological advance could have proceeded in step, resulting in a *truly* cultured and enlightened society? Or failing that, some in which the worst mistakes have been identified and corrected.

Most of us would like to believe that this has proved so among some of the other civilizations in the galaxy. The thought of encounters with such peoples is an exciting one. The words of the poet Tennyson, quoted at the beginning of this book, would then come true in a way and to an extent undreamed of by him. Harmony and trade with the star peoples—what an enthralling prospect for mankind! Once we sought the rich spices of the Indies and of remote Cathay, the precious metals of the Americas, the ivory of Africa. In future epochs we might secure products and ideas of immeasurably greater worth from the worlds of Epsilon Indi, Tau Ceti, Capella, or Altair.

Unfortunately, the other, less pleasant alternative remains. We all know of the elementary law of the jungle. Plants subscribe to it, for there are those that ruthlessly crowd out and "suffocate" others; insects prey on other insects; animals kill and eat other animals. Biped animals called humans do this too. Since they possess immeasurably higher intellectual and reasoning powers they generally do it for no reason at all—and because of these very powers have become extremely proficient at it! The law of the jungle has permeated all living forms on Earth. Perhaps it is, after all, just a basic biological law—the survival of the fittest. And since physical and chemical laws are valid throughout time and space it seems reasonable to expect biological ones to be equally valid. Seen in this light aliens do seem more likely to be predatory than benevolent.

At this point it is appropriate to dwell briefly on terrestrial reactions in the face of a full-scale armed invasion of our world by the forces of another. We must accept that the overwhelming military advantage would almost certainly lie with the invaders. Creatures able to cross the great gulfs of space are advanced beings by any standards. We must assume that their weapons and techniques would be equally sophisticated. No doubt the nations of Earth would put up a gallant and spirited defense but against such an adversary the issue could never really be in doubt. In the early years of this century the naval scene was revolutionized by the arrival of the British battleship *Dreadnought,* the first of the all-big-gun capital ships. At once all existing battleships were rendered obsolete, so much so in fact that a German admiral of the time dubbed them *"fünf minuten"* battleships—battleships likely to last only about five minues in combat with dreadnoughts. In a war with alien invaders Earth would probably find itself with *fünf minuten* armies, navies, and air forces! It would be a short business—though not mercifully so!

We have, of course, already seen it all in the pages of that epic novel by H. G. Wells, *The War of the Worlds,* and portrayed even more graphically as a motion picture a number of years ago. Modern terrestrial armories were simply useless against the remorseless invading Martians, and only terrestrial germs, to which the Martians were not immune, saved the peoples of Earth from total annihilation, perpetual servitude—or worse! At one point in the story one of the characters declares his intention of coming to terms with life on an Earth under total Martian domination. His philosophy is quite simple. Come what may, he intends to survive.

As a slave of the Martians at least he will be fed. (As readers of this story will discover, the Martians had a peculiar digestive system and a predilection for human blood, so it was more probable that they would have fed *on* him!) It is at best a doubtful philosophy, and in the circumstances, with all hope gone, death and merciful oblivion would probably be the lesser of two evils.

A subject race living under a competent, benign alien regime would certainly be a more endurable fate, however humiliating. Not that the peoples of Earth up to now have made an outstandingly good job of things on their own. On the contrary, they have succeeded in making one unholy mess—blatant, unashamed waste of valuable, nonrenewable resources, gross overpopulation, pollution of atmosphere, ocean, and land on an unprecedented and ever-increasing scale. The dismal list could go on and on. Whatever the future may hold for our civilization here on Earth, we could hardly complain if a superior galactic community took over and started to run it for us. The trouble is they would probably run it for themselves. Any benefit accruing to us might be more fortuitous than intentional.

It is also desirable to give some thought to the psychological effect on our kind from contact with an alien intelligence. Man has, over the centuries, created an image of himself as the pinnacle of creation. His ego might suffer disastrously were he ever to come face to face with representatives of a superior alien community which also regarded itself as "human." The resulting trauma might easily transform a superiority into an inferiority complex.

Such a contact could also change the entire fabric of our culture by interrupting or interfering with the logical process of thought and mental development. At least this is a view voiced in some quarters. It is interesting because it contrasts strongly with the more popular belief that, so long as the alien race were not hostile, nothing but good and progress could result from a fusion of ideas and ideals. The truth probably lies somewhere between these two extremes.

It is noteworthy that several eminent and highly respected scientists are now giving the matter of alien visitation a measure of thought. This is in itself something of a revolution and bears eloquent testimony to the changing outlook.

Zdenik Kopal of the University of Manchester, England, and an astronomer of international repute has recently augmented his advice about not answering the space-telephone if and when it rings. Convinced that intelligent beings from outer space are much more advanced than humans, he now expresses the belief that sooner or later an encounter with extra-terrestrials is, as he puts it, "inevitable." His precise conclusions are worth noting: "We might," he declares, "find ourselves in their test-tubes or other contraptions set up to investigate us, as we investigate insects or guinea-pigs!"

Sir Bernard Lovell, Director of the famous Jodrell Bank radio-observatory and one of the world's foremost radio astronomers, has also taken up the theme. In his presidential address to the British Association for the Advancement of Science, he warned of "hidden dangers" in the search for life on remote planets. "We must," he said, "regard life in outer space as a real and potential danger. You have only to think about the problems of diminishing resources here on Earth to realise that alien civilizations may be combing the galaxy looking for new resources or a new place to settle. They could want something we've got—and they could well have the ability to take it from us whether we liked it or not."

British Astronomer Royal, Sir Martin Ryle, has also added his voice and authority to the growing murmur. In a recent message to the International Astronomical Union he said, "There is a chance that someone is there!" When men of this caliber begin to sound a warning note, it is time for the world to pay heed.

And so, perhaps on a slightly less than cheerful note, we approach the end of our theme. Was Earth really visited by strange cosmic beings in the past? We can never be certain, but despite the terrifying immensity of space we see it at least as a possibility that cannot be ignored. Are we even now being visited or, what is more probable, being furtively and most effectively observed? While not desirous of giving undue impetus to the more enthusiastic, impulsive, and uninhibited of our "flying saucer" addicts, we feel bound to say that *certain* manifestations in our skies over the past few decades are both puzzling and intriguing. It is a little odd, one might even say disturbing, that these have coincided with mankind's recently acquired Promethean skills and his even more recently acquired ability to leave the environs of his own planet and walk upon the moon. If *our* immediate neighbors happened to be primitive cannibals, we would surely be troubled to learn they had acquired the ability to construct and use machine guns. Indeed, we would almost certainly feel disposed to do something about it—and the sooner the better. The parallel is obvious!

What then of the future? We have seen something of the possibilities. In this cosmos of star-laden galaxies it would be nothing short of incredible were Earth the only inhabited world. Undoubtedly there is life—other life—somewhere. But what of distance? Distance, as we have seen, can probably be destroyed, telescoped, given the appropriate technology. And to secure that appropriate technology the essential factor is time. In the case of solar systems evolving millennia ahead of our own this represents no problem. What else happens to worlds in time or more specifically to their populations? They expand. They not only expand; they expand both prodigiously and alarmingly. Just look at Earth. Better still, envisage it a century from now! A planet's people, like a nation's, needs *lebensraum*—living space. It must expand to somewhere. And where is somewhere? "Somewhere" must be a planet (or preferably planets) where atmospheric and other conditions are suitable, and since it is unlikely that these conditions will be duplicated in the same solar system it means neighboring solar systems. In due time it all happens again, and so more planets, more solar systems are needed. The real parameter is time. So what does all this add up to? It adds up quite clearly to a steadily advancing cosmic colonizing front—a great "bubble" in space slowly spreading all around its original nucleus. Will there be one solitary such "bubble"? This seems unlikely. There are probably several—which brings us immediately to the crux of the matter. How far from us now, at this very moment, is the perimeter of the nearest "bubble"? From its "frontier posts" may range the reconnaissance vessels, the eavesdropping probes, the alien cruisers, drawn like bees to honey by the crude nuclear detonations, and primitive attempts at space travel by the inhabitants of the third planet of a rather average yellow star!

Many times over the past few decades we have been told that either West or East was looking to its defenses. Perhaps all the while, even more so now, a *united* world should be looking to *its* defenses—such as they are! And should defense prove impossible, at least a policy, a plan, for coping with the unthinkable.

The dark, impalpable night sky is very peaceful, serene, and lovely. But is it quite as innocuous as it seems? It may be, could be, *just that little bit later than we think!*

# APPENDIX

Preparatory work for this book entailed perusal of a host of old records, reports, books, journals, newspapers, encyclopedias, and so on. These revealed very clearly that UFOs and odd manifestations in the skies are by no means a recent phenomenon. Indeed, references can be found dating back to the thirteenth century.

To give readers some idea of the scope, a short list of selected sightings is appended below. In the compilation of such a list certain criteria must obviously be observed, which is merely a formal way of saying that some attempt must be made to separate the sheep from the goats. The problem lies in knowing which are the sheep and which the goats!

Clear candidates for exclusion are items from the more remote historical past, up to and including the seventeenth century. During these times observation, superstition, astrology, theology, and black magic had a habit of becoming mixed. As a consequence, accuracy and authenticity are open to considerable doubt. From the early years of the eighteenth century to the start of the "flying saucer" era (1946), it is clear also that many alleged sightings can be attributed to unusual meteors, meteoric effects, auroral manifestations, or meteorological phenomena. These have therefore been eliminated also. Many more were vague either with respect to description, locality, time, etc. These too have been excluded.

The list ends at 1945. Thereafter reports of sightings became so numerous that a list, even a severely edited one, would virtually constitute a book in its own right. No absolute guarantee can be given for the authenticity of the items in this list. Most, however, emanate from fairly reliable sources. In view of the number and frequency of sightings over the years and the fact that few parts of the world appear to have remained unaffected it is difficult to avoid the feeling that for a long time now something rather inexplicable has been happening.

| DATE | LOCALITY | DESCRIPTION |
|---|---|---|
| 1752, April 15 | Stavanger, Norway | Bright aerial object having octagonal shape. |
| 1762, August 9 | Basle, Switzerland | Huge, dark object said to be "spindle-shaped" and surrounded by a glowing outer ring seen crossing in front of the sun's disc by two Swiss astronomers. |
| 1779, June 7 | Boulogne, France | Formation of "glowing discs" passes over city. |
| 1808, October 12 | Piedmont, Italy | Formation of "luminous discs" observed. |
| 1816, October? | Edinburgh, Scotland | Large, luminous, crescent-shaped object seen over city by many people. |
| 1820, September 7 | Embrun, France | Formation of flying objects crosses the town. |
| 1821, November 22 | English Channel | Luminous disc observed from several ships in the English Channel. |
| 1826, April 1 | Saarbrücken, France | Gray, torpedo-shaped object seen in sky. |
| 1831, November 29 | Thuringia, Germany | Brilliant luminous disc seen in sky. |
| 1833, November 13 | Niagara Falls, USA | Luminous flying object crosses city. |
| 1836, January 12 | Cherbourg, France | Gleaming aerial vessel seen over city. |
| 1845, May 11 | Naples, Italy | Astronomer at local observatory reports number of luminous discs leaving trails. |

| DATE | LOCALITY | DESCRIPTION |
| --- | --- | --- |
| 1846, October 26 | Lowell, Mass., USA | "Luminous flying disc" reported over the area. |
| 1847, March 19 | London, England | "Spherical craft" seen rising vertically through the clouds by several witnesses. |
| 1848, September 19 | Inverness, Scotland | Two large, brilliant aerial objects seen over town. |
| 1850, June 6 | Côte d'Azur, France | Red sphere seen in sky. |
| 1853, October 26 | Ragusa, Sicily | Brilliant disc observed moving from east to west during early hours of morning. |
| 1855, August 11 | Sussex, England | Glowing, circular object "with spokes like a wheel" crosses sky. Visible for an hour. |
| 1856, April 6 | Colmar, France | Dark "aerial torpedo" seen in sky. One end round, the other pointed. Said to have emitted low-pitched sound. |
| 1863, April 27 | Zurich, Switzerland | Zurich Observatory reports "aerial discs" emitting "whining sound." |
| 1864, October 10 | Paris, France | Noted French astronomer Leverrier reports appearance of luminous tubular object over city. |
| 1868, June 8 | Oxford, England | Radcliffe Observatory reports four-minute appearance of luminous flying object. |
| 1871, August 1 | Marseilles, France | Large red disc hovers over city for almost ten minutes. |
| 1874, July 6 | Oaxaca, Mexico | Large conical object observed over the area. |
| 1877, September 7 | Indiana, USA | Dark objects seen crossing sky emitting flashes at four-second intervals. |

| DATE | LOCALITY | DESCRIPTION |
|---|---|---|
| 1879, May 15 | Persian Gulf | British warship reports two huge "spinning aerial wheels" just above surface of the sea. |
| 1880, July 30 | St. Petersburg, Russia | Large airborne spherical object accompanied by two lesser spheres. |
| 1882, November 17 | Greenwich, England | Vast green disc observed from Greenwich Observatory. Mottled appearance. Reported also from Continent. |
| 1883, November 5 | Santiago, Chile | Pulsating disc-like object passes over city. |
| 1884, July 3 | New York, USA | Bright spherical craft with dark markings reported from many parts of New York State. |
| 1885, February 24 | Pacific Ocean, northwest of Hawaii | Master of a freighter in the area reports fall of "a huge fiery mass" into sea near ship. |
| 1888, November 3 | Reading, England | Aerial vessel passes over county of Berkshire. Animals allegedly affected by sound waves over a considerable area. |
| 1893, May 25 | Sea of Japan | Two British warships report formation of discs heading north and emitting smoke trails. |
| 1897, April 14 | Kansas City, USA | Torpedo-shaped object observed with downward shining searchlight. |
| 1899, November 1 | Dumfries, Scotland | Two circular black objects cross sky during daylight. One said to be "circling the other." |
| 1901, April 4 | Persian Gulf | Master of freighter reports |

| DATE | LOCALITY | DESCRIPTION |
| --- | --- | --- |
| | | "revolving luminous wheels" near surface of the sea. |
| 1905, March 29 | Cardiff, Wales | Luminous, vertical tube-shaped object observed in sky. |
| 1908, December 1 | Sofia, Bulgaria | Bright spherical object flies slowly across city. |
| 1909, March 17 | Peterborough, England | Police report passage of "aerial object" across town. |
| 1910, May 4 | Cirnovti, Russia | Mathematics professor reports appearance of flying object with estimated diameter of 100 meters. |
| 1913, January 31 | South Wales | "Tube-shaped object" with lights seen from many parts of South Wales. |
| 1915, July 19 | West Virginia, USA | Luminous cylindrical object seen from several parts of the state. |
| 1923, January 18 | Southeast Scotland | Reports of strange aerial craft, torpedo-shaped, faint blue glow. |
| 1928, July 7 | Piatraolt, Rumania | Luminous cylindrical object seen by many witnesses; flying at high speed in easterly direction but emitting no sound. |
| 1936, July 24 | Peebles, Scotland | Faintly glowing cylindrical object crosses night sky. |
| 1938, February 28 | Newcastle-on-Tyne, England | Circular aerial craft sighted heading north. Flickering light. |
| 1943, September 7 | Pushkino, USSR | Disc-shaped object observed high above aerial combat between Russian and German aircraft. |
| 1945, August 9 | Vulcanesti, Rumania | Large red object crosses sky. Seen by passengers and crew of Bucharest express. |

# REFERENCES

*Chapter 2*

1. M. H. Briggs, "Detection of Planets at Interstellar Distances," *Journal of the British Interplanetary Society* 17 (1959): p. 59.
2. D. A. Lunan, "Space Probe from Epsilon Boötis," *Spaceflight* 15 (London: April 1973): pp. 122-131.

*Chapter 3*

1. William Bonnor, *The Mystery of the Expanding Universe* (New York: Macmillan, 1964): p. 81.
2. Richard P. Feynman, *Feynman Lectures on Physics,* Vol. 2 (Reading, Mass.: Addison-Wesley, 1964): pp. 42.2-42.5.
3. Don Albers, "The Meaning of Curved Space," *Mercury* (San Francisco: Astronomical Society of the Pacific, July-August 1975): pp. 16-19.

*Chapter 4*

1. John Taylor, *Black Holes* (London: Souvenir Press, 1973; New York: Random House, 1974).
2. W. J. Kaufman III, "Pathways Through the Universe–Black Holes, Worm Holes and White Holes," *Mercury* (San Francisco: Astronomical Society of the Pacific, May-June 1974): pp. 26-33.
3. Ibid: pp. 26-33.

*Chapter 5*

1. L. Marder, *Time and the Space-Traveller* (London: Fakenham, 1971; Philadelphia: University of Pennsylvania Press, 1974).

2. J. W. Morgan, "Superrelativistic Interstellar Flight," *Spaceflight* 17 (London: July, 1973): p. 252.

*Chapter 6*

1. O. M. Bilanuick et al., "Particles Beyond the Light Barrier," *Physics Today* (New York: American Institute of Physics, May 1969): p. 143.
2. R. G. Newton, "Particles That Travel Faster Than Light," *Science* (Washington, D.C.: American Association for the Advancement of Science, 3925): p. 1569.
3. G. Feinberg, "Possibility of Faster Than Light Travel," *Physical Review* 1959 (London: 1970): p. 1089.
4. J. W. Morgan, "Superrelativistic Interstellar Flight," *Spaceflight* 17 (London: July 1973): p. 252.
5. G. Feinberg, "Possibility of Faster than Light Travel."
6. J. W. Morgan, "Superrelativistic Interstellar Flight," *Spaceflight* 17 (London: July 1973): p. 252.

*Chapter 7*

1. James Strong, "Trans-Stellar Navigation," *Spaceflight* 17 (London: July 1971): pp. 252-255.
2. E. Sanger, "Some Optical and Kinematic Effects in Interstellar Astronautics," *Journal of the British Interplanetary Society* 18 (1961-62): pp. 273-276.
3. J. W. Macvey, *Journey to Alpha Centauri* (New York: Macmillan, 1975), Chapter 13.
4. Strong, "Trans-Stellar Navigation," *Spaceflight* 13 (London: July 1971): pp. 252-255.

*Chapter 8*

1. P. M Molton, "Terrestrial Biochemistry in Perspective: Some Other Possibilities," *Spaceflight* 15 (London· April 1973): p. 139.
2. R. A. Horne, "On the Unlikelihood of Non-Aqueous Biosystems," *Space Life Science* 3 (London: 1971): pp. 34-41.
3. G. C. Pimental et al., "Exotic Biochemistry in Exobiology," (Washington, D.C.: National Academy of Science, 1966) No. 1296: pp. 243-251.

4. L. H. Aller, "Some Aspects of the Abundance Problem in Planetary Nebulae," *Journal of the Astronomical Society of the Pacific* 76 (San Francisco: 1964): pp. 279-288.
5. L. F. Herzog, "Determining the Composition and History of the Solar System," *Analytical Chemistry in Space,* R. E. Nainerdi, ed. (London and Elmsford, N.Y.: Pergamon Press, 1970): p. 8.
6. P. Molton, "Exobiology, Jupiter and Life," *Spaceflight* 14 (London: June 1972): pp. 220-223.
7. P. M. Molton, "Terrestrial Biochemistry in Perspective."
8. Ibid.
9. Aller, "Some Aspects of the Abundance Problem."
10. Herzog, "Determining the Composition and History."

*Chapter 9*

1. G. V. Foster, "Non-human Artifacts in the Solar System," *Spaceflight* 14 (London: December 1972): pp. 447-453.
2. P. A. Sneath, *Planets and Life* (London: Thames & Hudson, 1969; New York: Funk & Wagnalls, 1970).
3. K. A. Ehricke, "Astrogenic Environments," *Spaceflight* 14 (London: 1972): p. 1.
4. C. W. Anderson, "A Relic Interstellar Corner Reflector in the Solar System," *Mercury* 3 (Astronomical Society of the Pacific, September-October 1974): pp. 2-3.
5. C. E. S. Horsford, "A British Code of Space Law," *Spaceflight* 5 (London: 1963): p.2.
6. D. A. Lunan, "A Space Probe from Epsilon Boötis," *Spaceflight* 15 (London: April 1973): pp. 122-131.

*Chapter 12*

1. A. E. Roy, "Do You Believe in Flying Saucers?" *Glasgow Herald,* November 15, 1974.
2. Peter Kolosimo, *Not of This World* trans. A. D. Hills (London: Souvenir Press, 1974; New York: Bantam Books, 1973).
3. Ibid.
4. *Times,* London: June 22, 1944.
5. D. Brewster, British Association for the Advancement of Science, Report (1848-51).
6. B. Belitsky, "Reflections on Ceti," *Spaceflight* 15 (London: July 1973): p. 255.

7. S. Milton, "Status of Science in Prehistory," *New Scientist* 56 (London: December 14, 1972): pp. 636-637.
8. Ibid.
9. Erich von Daniken, *Chariots of the Gods?* (London: Souvenir Press, 1969; New York: G. P. Putnam's Sons, 1970).
10. D. A. Lunan, *Man and the Stars* (London: Souvenir Press, 1974): p. 296.
11. Erich von Däniken, *Chariots of the Gods?*
12. Lunan, *Man and the Stars.*
13. S. Milton, "Megalithic Moonwatchers of Europe," *New Scientist* 54 (London: April 13, 1972): pp. 60-62.
14. Alexander Thom, *Megalithic Lunar Observatories* (London and New York: Oxford University Press, 1971).
15. Ibid.
16. Alexander Thom, "Megalithic Observatories," *Journal of the History of Astronomy* 2 (1971): p. 147.
17. Milton, "Megalithic Moonwatchers of Europe."
18. Milton, "Status of Science in Prehistory."
19. Ibid.
20. Ibid.

*Chapter 14*

1. J. W. Macvey, "Missiles from Space," *Spaceflight* 10 (London: February 1968): pp. 46-48.
2. H. P. Hollis, *Journal of British Astronomical Association* 18 (London: September 1908): p. 354.
3. No author named, "Siberian Meteor of June 30, 1908," *Nature* 127 (London: May 9, 1931): p. 719
4. C. Cowan, C. R. Atluri, W. F. Libb, "Possible Anti-Matter Content of Tunguska Meteor of 1908," *Nature* 206 (London: May 29, 1965): pp. 861-865.
5. Ibid, p. 861.
6. Ibid, p. 861.
7. Ibid, p. 861.
8. Ibid, p. 861.
9. Ibid, p. 861.
10. E. L. Krinov, *Solar System* (University of Chicago Press, 1963).
11. F. J. Whipple, "Great Siberian Meteor," *Quarterly Journal of Royal Meteorological Society* 56 (London: 1930): p. 287.
12. V. G. Fesenkov, "On the Cometary Nature of the Tungus Mete-

orite," *Astronomicheskii Zhurnal* 38 (Moscow: April 1961): pp. 577-592.
13. Ibid, pp. 577-592.
14. V. G. Fesenkov, "The Nature of the Tunguskan Meteorite," *Meteoritika* 20 (Moscow: 1961): pp. 27-31.
15. A. Zolotov, "Estimation of the Parameters of the Tungus Meteorite Based on New Data," *Soviet Physics Doklady* 12 (Moscow: 1967): p. 108.
16. E. L. Krinov, *Giant Meteorites* (London: Pergamon Press, 1966).
17. V. G. Fesenkov, "The Air Wave Caused by the Fall of the Tunguska Meteorite," *Meteoritika* 17 (Moscow: 1959): pp. 2-7.
18. A. A. Jackson, M. P. Ryan, "Was the Tunguska Event Due to a Black Hole?" *Nature* 245 (London: September 1973): p. 88.
19. Ibid, p. 88.
20. D. M. Eardley, "Tungus Black Hole?" *Nature* 245 (London: October 1973): p. 397.
21. G. L. Wick, J. D. Isaacs, "Tunguska Meteorite," *Nature* 247 (London: January 1974): p. 139.

*Chapter 15*

1. British Ministry of Defence Report, 1968.
2. Barry Goldwater, "I Believe Earth has been Visited by Creatures from Outer Space," *National Enquirer,* January 6, 1974.
3. John South, "Frightened Crowd Sees a Hissing, Glowing U.F.O. in Chattanooga, Tenn.," *National Enquirer,* January 13, 1974.
4. Peter Gwynne, "Flying Watergate Saucers," *New Scientist* 60 (London: November 1, 1973): p. 35.
5. Ibid, p. 35.
6. Ibid, p. 35.
7. Ibid, p. 35.
8. Ibid, p. 35.
9. Barry Goldwater, (As per No. 2 above).
10. R. Bryant, "The Night President Carter Saw a Flying Saucer," *Week End* (London: February 23, 1977).
11. Ibid.
12. Ibid.
13. Ibid.

# INDEX